Coal

An Anthology of Mining

Coal

An Anthology of Mining

Edited by
TONY CURTIS

Foreword by
Tony Benn MP

seren is the book imprint of
Poetry Wales Press Ltd,
Wyndham Street, Bridgend,
CF31 1EF, Wales

Editorial & Introduction © Tony Curtis, 1997
Foreword © Tony Benn, 1997
For individual contributions see Acknowledgements pages

ISBN: 1-85411-188-4

A CIP record for this title is available from
the British Library

*The publisher acknowledges the financial support
of the Arts Council of Wales*

Cover illustration: Josef Herman

Printed in Plantin by
Creative Print and Design, Ebbw Vale

Contents

The Working Day

The Women

Strife: Strikes, Unions and Politics

And Gold On My Neck the Sun

Foreword

This book celebrates coal-mining and that very phrase conjures up such a huge range of memories and events, and draws attention to the key role that coal has played in our history. The industrial revolution was built on coal, and coal-fired ships carried manufactures around the world, while coal-fired warships protected these trade routes. No wonder coal was known as 'King'.

But it is not the material but the people who mined it who made its history so important. At the beginning of the century three quarters of a million men went underground every day to dig the nation's coal, and the whole mining culture has been immensely important to the country in so many ways.

Unlike those in the City who earn their living by outsmarting their neighbours, miners survive by relying totally upon those working with them at the coal-face, sharing the adversity of explosions, roof falls or other possible calamities. Miners — like fishermen and farmers — have evolved a special type of tribal solidarity because they are up against a nature which can be very harsh.

Mining communities have always been special. The loyalty and trust they breed are expressed in their local institutions and through political action. Miners have played a huge role in shaping our society, most famously Keir Hardie, A.J. Cook, Nye Bevan and Arthur Scargill. And mining communities are not just about the men themselves. The role of women in the mining industry is also fascinating; when the going has got rough the women have emerged as people of great imagination and strength. In earlier times, of course, they worked down the pits themselves.

When Mrs Thatcher took on the NUM in 1984-85 it was a conscious declaration of civil war against the miners and what they stand for. If the Labour movement had supported the miners, they would have won and changed the history of the years that followed. Yet one day we shall have to go back to coal, for there are a thousand years reserves still underground and it is Britain's greatest mineral resource.

It is no surprise that out of this long and thrilling story of human courage and endeavour so many banners and poems and pictures and stories and songs have emerged along with the recollections of those who took part. Some are recorded in this book, which may tell us more about the country in which we live than the boring gossip about 'people at the top' who hover about but contribute nothing to the development of national life.

Tony Benn

Introduction

This is the first major collection of writing about the coal industry
and the communities which developed around it across the British
Isles. From the observations of George Owen of Henllys in 1603
to the writings of Barry Hines and Duncan Bush witnessing the
death-throes of the industry in the 1980s, this book shows that
coal mining and the experiences of its communities have inspired
some of our most memorable writing. I have brought together a
wide range of fiction and poetry and song from four centuries and
the four corners of these islands. D.H. Lawrence, George Orwell,
Gwyn Thomas, Dennis Potter, Idris Davies, Dannie Abse, Philip
Larkin and many other significant writers are represented alongside ballads, folk-verse, autobiography and fictionalised remem-
brances. The book is organised into six sections which focus on
the central experiences of the men, women and children whose
lives revolved around mining.

Coal has been the major factor in British history for over three
hundred years. It fuelled the Industrial Revolution, drove the
British Empire and its armed forces for a century, and it affected
great sections of the population in the way people lived, and where
they lived, to a greater extent than any other factor. Its social
impact remains: I teach now in the University of Glamorgan, in
Treforest, in a building established in 1913 as the site of the South
Wales and Monmouth School of Mines. Despite 14,000 students
from around the world and a brand new logo, some older local
people still refer to it as 'the school of mines'. The river Taff which
flows past is now clean enough to take those salmon making their
way to spawn in the streams of Breconshire. Over fifty mines have
closed in south Wales. But just above the valleys communities
which the University also serves, the Tower Colliery at Hirwaun,
the last mine in the coal field, has announced another year of profit
since its miners bought it. Against all predictions they are mining
good coal for good wages, and a share of the profits.

I come from rural west Wales, beyond the coalfield as we know
it, but when we tried to dig our garden in Pembrokeshire the soil
contained dark shale, once on its way to being coal. In the early
1970s I taught in a comprehensive school in Maltby in Yorkshire.
Many of the families in that village had moved there, or stayed
there because of the pit. Maltby, Mexborough and the other mines
in the Doncaster area still dominated the lives and jobs of the local
population. When Barry Hines came as a guest to talk about his
novel *Kestrel for a Knave* and the film *Kes*, it was clear to both of

us that his audience contained many Billy Caspers and their sisters; these kids were watching their own lives go by. They were either cowed or resilient; the boys knew that the pit was their likely future, with luck, but they also loved the casual farm work they did, harvesting, beating for the shoot, and training ferrets. There, as in many other coalfields, the darkness of the work underground was tempered by the surrounding countryside. The neighbouring pit village of Mexborough has produced poets Ted Hughes and Harold Massingham, two celebrators of nature. In the 1984 strike the police station at Maltby was taken over by protesters; the Billy Caspers were fighting back, in vain.

The town of Barry where I now live is no older than many American cities. Its grid-lines of terraced houses grew into the profitable dream of David Davies of Llandinam and the Earl of Plymouth who, in 1884, succeeded in obtaining Parliament's permission to create a port. So much coal was coming down from the south Wales valleys that the docks of Swansea, Newport and Cardiff could not cope. Barry docks, opened in 1889, was by 1913 the biggest coal-exporting port in the world. And in the next four decades the other, sandy side of Barry Island was the day tripper resort where miners could give their families a day at the sea. When we first moved to Barry in 1974 we would walk our son's pram along the dockside under the lines of rusting coal-loading hoists, their giant legs striding out from a web of railway lines. A couple were still in use and from a safe distance we would watch the coal wagons picked up and inverted over the holds of ships like parts of a Hornby train set. Now that dock is landscaped to accept the proposed housing and marina. That's progress — but if you retrace the missing railway tracks back to the valleys of Glamorgan and Gwent you find wasted communities, retired or redundant miners, perhaps silicotic, watching their grandchildren's generation eroded by unemployment and drugs. There are no simple equations, only shifting experiences: naturally, the writings in this anthology are at times polemical, at times lyrical.

Coal was under us all the time. In Wales, south and north east, in Scotland, the north of England and the Midlands, from Pembrokeshire to Kent, from Lancashire to the Forest of Dean, in Yorkshire, Durham and Nottingham generations have lived from and for coal; coal lit their fires and their politics; coal crushed men and fused individuals into strong families and communities. Coal made fortunes and could bring down a government; coal gave birth to towns and ports, to railways and roads. And when a government closed mines, it strangled the life from the valleys and towns it had tolerated out of necessity. The communities that died government knew little of and cared nothing for. I hope this anthology helps to record and celebrate those lives lived and lost.

Tony Curtis

Editor's Acknowledgements

Thanks are due to Sheenagh Pugh, Mac Daly, Malcolm Lewis, the National Library of Scotland, Mick, Simon, and Claire at Seren and the Regional Research Fund of the University of Glamorgan, all of whom were generous in their time and support for this project. However, the editorial decisions were my own.

I am also indebted to Tony Benn for his Foreword, and to Josef Herman R.A. for the drawings of the miners of Ystradgynlais on the cover and inside the book, made in the 1940s and 50s.

The Fair Country

Wilfred Owen
Miners

There was a whispering in my hearth,
  A sigh of the coal,
Grown wistful of a former earth
  It might recall.

I listened for a tale of leaves
  And smothered ferns;
Frond-forests: and the low, sly lives
  Before the fauns.

My fire might show steam-phantoms simmer
  From Time's old cauldron,
Before the birds made nests in summer,
  Or men had children.

But the coals were murmuring of their mine,
  And moans down there
Of boys that slept wry sleep, and men
  Writhing for air.

And I saw white bodies in the cinder-shard,
  Bones without number;
Many the muscled bodies charred;
  And few remember.

I thought of some who worked dark pits
  Of war, and died
Digging the rock where Death reputes
  Peace lies indeed.

Comforted years will sit soft-chaired
  In rooms of amber;
The years will stretch their hands, well-cheered
  By our lives' ember.

The centuries will burn rich loads
  With which we groaned,
Whose warmth shall lull their dreaming lids
  While songs are crooned.
But they will not dream of us poor lads
  Left in the ground.

Richard Llewellyn
The Slag Heap

The slag heap is moving again.

I can hear it whispering to itself, and as it whispers, the walls of this brave little house are girding themselves to withstand the assault. For months, more than I ever thought it would have the courage to withstand, that great mound has borne down upon these walls, this roof. And for those months the great bully has been beaten, for in my father's day men built well for they were craftsmen. Stout beams, honest blocks, good work, and love for the job, all that is in this house.

But the slag heap moves, pressing on, down and down, over and all round this house which was my father's and my mother's and now is mine. Soon, perhaps in an hour, the house will be buried, and the slag heap will stretch from the top of the mountain right down to the river in the valley. Poor river, how beautiful you were, how gay your song, how clear your green waters, how you enjoyed your place among the sleepy rocks.

Idris Davies
O What Can You Give Me?

O what can you give me?
Say the sad bells of Rhymney.

Is there hope for the future?
Cry the brown bells of Merthyr.

Who made the mineowner?
Say the black bells of Rhondda.

And who robbed the miner?
Cry the grim bells of Blaina.

They will plunder willy-nilly,
Say the bells of Caerphilly.

They have fangs, they have teeth!
Shout the loud bells of Neath.

To the south, things are sullen,
Say the pink bells of Brecon.

Even God is uneasy,
Say the moist bells of Swansea.

Put the vandals in court!
Cry the bells of Newport.

And would be well if — if — if —
Say the green bells of Cardiff.

Why so worried, sisters, why?
Sing the silver bells of Wye.

D.H. Lawrence
Living in the Square

Even fifty years ago the squares were unpopular. It was 'common'
to live in the Square. It was a little less common to live in the
Breach, which consisted of six blocks of rather more pretentious
dwellings erected by the company in the valley below, two rows
of three blocks, with an alley between. And it was most 'common',
most degraded of all to live in Dakins Row, two rows of the old
dwellings, very old, black four-roomed little places, that stood on
the hill again, not far from the Square.

So the place started. Down the steep street between the squares,
Scargill Street, the Wesleyans' chapel was put up, and I was born
in the little corner shop just above. Across the other side of the
Square the miners themselves built the big, barn-like Primitive
Methodist chapel. Along the hill-top ran the Nottingham Road,
with its scrappy, ugly mid-Victorian shops. The little market-
place, with a superb outlook, ended the village on the Derbyshire
side, and was just here left bare, with the Sun Inn on one side, the
chemist across, with the gilt pestle-and-mortar, and a shop at the
other corner, the corner of Alfreton Road and Nottingham Road.

In this queer jumble of the old England and the new, I came
into consciousness. As I remember, little local speculators already
began to straggle dwellings in rows, always in rows, across the
fields nasty: red-brick, flat-faced dwellings with dark slate roofs.
The bay-window period only began when I was a child. But most
of the country was untouched.

There must be three or four hundred company houses in the
squares and the streets that surround the squares, like a great
barracks wall. There must be sixty or eighty company houses in

the Breach. The old Dakins Row will have thirty to forty little
holes. Then counting the old cottages and rows left with their old
gardens down the lanes and along the twitchells, and even in the
midst of Nottingham Road itself, there were houses enough for
the population, there was no need for much building. And not
much building went on when I was small.

We lived in the Breach, in a corner house. A field-path came
down under a great hawthorn hedge. On the other side was the
brook, with the old sheep-bridge going over into the meadows.
The hawthorn hedge by the brook had grown tall as tall trees, and
we used to bathe from there in the dipping-hole, where the sheep
were dipped, just near the fall from the old mill-dam, where the
water rushed. The mill only ceased grinding the local corn when
I was a child. And my father, who always worked in Brinsley pit,
and who always got up at five o'clock, if not at four, would set off
in the dawn across the fields at Coney Grey, and hunt for
mushrooms in the long grass, or perhaps pick up a skulking rabbit,
which he would bring home at evening inside the lining of his
pit-coat.

So that the life was a curious cross between industrialism and
the old agricultural England of Shakespeare and Milton and
Fielding and George Eliot. The dialect was broad Derbyshire, and
always 'thee' and 'thou'. The people lived almost entirely by
instinct, men of my father's age could not really read. And the pit
did not mechanize men. On the contrary. Under the butty system,
the miners worked underground as a sort of intimate community,
they knew each other practically naked, and with curious close
intimacy, and the darkness and the underground remoteness of
the pit 'stall', and the continual presence of danger, made the
physical, instinctive, and intuitional contact between men very
highly developed, a contact almost as close as touch, very real and
very powerful. This physical awareness and intimate *togetherness*
was at its strongest down pit. When the men came up into the
light, they blinked. They had, in a measure, to change their flow.
Nevertheless, they brought with them above ground the curious
dark intimacy of the mine, the naked sort of contact, and if I think
of my childhood, it is always as if there was a lustrous sort of inner
darkness, like the gloss of coal, in which we moved and had our
real being. My father loved the pit. He was hurt badly, more
than once, but he would never stay away. He loved the contact,
the intimacy, as men in the war loved the intense male comradeship of the dark days. They did not know what they had lost till
they lost it. And I think it is the same with the young colliers of to-day.

Now the colliers had also an instinct of beauty. The colliers' wives had not. The colliers were deeply alive, instinctively. But they had no daytime ambition, and no daytime intellect. They avoided, really, the rational aspect of life. They preferred to take life instinctively and intuitively. They didn't even care very profoundly about wages. It was the women, naturally, who nagged on this score. There was a big discrepancy, when I was a boy, between the collier who saw, at the best, only a brief few hours of daylight — often no daylight at all during the winter weeks — and the collier's wife, who had all the day to herself when the man was down pit.

The great fallacy is, to pity the man. He didn't dream of pitying himself, till agitators and sentimentalists taught him to. He was happy: or more than happy, he was fulfilled. Or he was fulfilled on the receptive side, not on the expressive. The collier went to the pub and drank in order to continue his intimacy with his mates. They talked endlessly but it was rather of wonders and marvels, even in politics, than of facts. It was hard facts, in the shape of wife, money, and nagging home necessities, which they fled away from, out of the house to the pub, and out of the house to the pit.

The collier fled out of the house as soon as he could, away from the nagging materialism of the woman. With the women it was always: This is broken, now you've got to mend it! or else: We want this, that, and the other, and where is the money coming from? The collier didn't know and didn't care very deeply — his life was otherwise. So he escaped. He roved the countryside with his dog, prowling for a rabbit, for nests, for mushrooms, anything. He loved the countryside, just the indiscriminating feel of it. Or he loved just to sit on his heels and watch — anything or nothing. He was not intellectually interested. Life for him did not consist in facts, but in a flow. Very often, he loved his garden. And very often he had a genuine love of the beauty of flowers. I have known it often and often, in colliers.

Lewis Jones
An Evening in Cwmardy

Len's eyes widened at sight of the forbidding black power-houses and the lake of feeding water near them. A score of pipes with tiny holes intersected this latter and sent bubbling sprays of boiling water into the air, where each became a miniature rainbow before falling back into the lake.

Len could not take his eyes from this effervescent, sparkling cascade. 'Is all the colours in the world by there, Dad?' he asked wonderingly.

Big Jim coughed and blew his nose. 'Ay, I think they all be there, boy bach,' he answered doubtfully. 'But,' he added, 'you did ought to see it first thing in the morning when the sun be rising. Ah, that be a sight for sore eyes.'

Len thought a moment before saying, 'When I grow up and start to work I will be able to see it in the morning, 'on't I, Dad?'

'Ay, ay, my boy,' answered Big Jim; adding under his breath, 'When you start to work you 'on't want to see it.' Len did not hear this.

The thought of working in the pit sent ripples through his flesh and made him anxious to grow up quickly. He looked at the trellised iron-work of the pit-head frames, to which the power houses made such an appropriate background, and sighed long-ingly. The gleaming rail-tracks which radiated from the pit-head down the valley towards the invisible sea seemed, in his imagina-tive eyes, to be like veins of quickly coursing blood. Big Jim broke into his, thoughts with the command, 'Follow me, and be careful how you do tread.'

Alun Lewis
The Mountain over Aberdare

From this high quarried ledge I see
The place for which the Quakers once
Collected clothes, my father's home,
Our stubborn bankrupt village sprawled
In jaded dusk beneath its nameless hills;
The drab streets strung across the cwm
Derelict workings, tips of slag
The gospellers and gamblers use
And children scrutting for the coal
That winter dole cannot purvey;
Allotments where the collier digs
While engines hack the coal within his brain;
Grey Hebron in a rigid cramp,
White cheap-jack cinema, the church
Stretched like a sow beside the stream;
And mourners in their Sunday best

Holding a tiny funeral, singing hymns
That drift insidious as the rain
Which rises from the steaming fields
And swathes about the skyline crags
Till all the upland gorse is drenched
And all the creaking mountain gates
Drip brittle tears of crystal peace;
And in a curtained parlour women hug
Huge grief, and anger against God.

But now the dusk, more charitable than Quakers,
Veils the cracked cottages with drifting may
And rubs the hard day off the slate.
The colliers squatting on the ashtip
Listen to one who holds them still with tales
While that white frock that floats down the dark alley
Looks just like Christ; and in the lane
The clink of coins among the gamblers
Suggests the thirty pieces of silver.

I watch the clouded years
Rune the rough foreheads of these moody hills,
This wet evening, in a lost age.

Duncan Bush
The Head of the Valley

The road climbs slowly, wooded hills on either side, the odd field
of sheep, I slow to look as I pass a pit on the left: a yard full of
coaldust and yellow plant, a chained gate, half a dozen men
around a fire in an oildrum brazier. A hand-painted sign is
propped against the link fence:

NUM Oficial Picket

The second *f* is corrected in above in a narrowed stroke. And I
remember the miners are out on strike, have been for a while now.
There's been a lot about it on the TV news.
   In a circle playing cards, they stop and look up as I slow, holding
it long enough to be a stare. Woollybacks. Up here I don't suppose
they've seen a car like this before, except in films.

A mile on I have to slow again for a village, or at least some rows of terraced houses. The front door opening straight onto the pavement, the odd old lady sitting in the doorway in a brought chair, watching things. Kids playing in the street.

Then, right at the head of the valley, is a place I remember, and I know I'm almost at the end of the road. It's only one more dirty-looking, Welsh, played-out little village, don't even ask me to pronounce the fucking name of it. Just another place that grew up around a hole in the ground twenty miles from nowhere, a couple of dozen or so houses, every one complete with outside toilets, and views of half-grassed slagheaps if you take a glance out of the wrong window.

But what I remember from last time is when you take the sharp left out of the place and swing round the corner and across the railway bridge, and suddenly for the first time, right across the next valley, you see the mountains.

And on a clear day like it is again today it's like Wyoming must be or Montana, like a background you'd see in a film. Or some Shangri-la, another mountain land painted on glass. And I slowed the car right down last time to look at it, and I do again now, and it gives me the same lift this time as last. Because it's suddenly like you're not in Wales any more. You're in America.

So drive. Leaving the railway-sidings and the massive coalyard dunes and another chained gate with a half-dozen men who again immediately stop playing cards in front of it to stare me past, the stranger in the two-tone Yankee car.

Then I rattle across a cattle-grid (except that up here you'd have to say it was a sheep-grid). And, on the other side of that, I pass the roadside sign that tells you:

Powys

And there are no pits here of course, no slagheaps. Only those mountains. And the Brecon Beacons National Park. Because what starts here is a Conservation area. It's as if you pass that sign out of West Glamorgan into Powys and the coalfield stops right at the boundary, either because there's no coal in the ground in Powys or they wouldn't let you make the mess it takes to get it out.

Harold Massingham
Frost-Gods

An air-berg's exploded!
                    All's
Frost-flak over these collieries, this stubbled corn. —

Tablets dissolve like this,
The only movement: microbe, atom, visible air,
A glistering, getting nowhere.

It is a place I can stand
More than most.
                    I look it over, find
Grey cress on the pond,

Streaked water in ruts like inlaid lead.
I see how progress re-settled Denaby,
Slag-heap, wheel, chimney, a mile away —

And here, how thistles have settled what they can.
Behind me, the river
Rattles through its weir like marbles,

The canal's peaceful and oil-skinned:
                                    and ahead,
The hump of Old Denaby
Carries tree-boles like football-studs,

Like warts.
                — I look it over, warming
To that grim aura: till I stiffen,
By frost, frost-stroke

That petrifies!
                    Heritage, home-truth,
Some biotic thing, a virus
Out of control!
                    Whose Breathings are they,

Who visit, looming
Like Histories from blizzardous homes, clan-camps of
    January
To a stand-still here?

All's apparition,
Settled by Danish brain,
Of stag wood, fiord and conifer, eider-skies

Of Odin.
            Now in their shimmer of stoat-fur,
Clinging mould of fridges,
I see them through staring blains —

Jacks-of-all-Winters, born to embroider,
Born to age meadows,
Scare geese from whitening marshes,

Ford the North Sea and haze
Into England.
            There should be ram-horns, a-hoyes
    of gannets,
But their adventures are silent:

They're not civil,
Not righteous or malevolent,
But energied with Reverie,

With Operation —
So build fires, invent embodiments and names
For shock meetings

Such as this, as Northness stands
Holding its thistling easterlies in one inescapable
Hush!
            — And naming them,

Begin worship,
By exhilaration and the brain's flickering charms.

— Ten seconds, and Imagination's
On hard ground: like a shrub, it has put forth
Berries.
            It's humours

Have adored. —
                    Returning to narrow streets,
Their windows draped with poor gauze,
And my father before his fire,

I leave Old Denaby waiting for snow,
Its thistles hardening to the root, its hump
Arching like a white stoat.
                                                    And I fancy

The frost's blur of grey-midge over that pond,
And Denaby's chimneys smoking gently like snuffed
candles.

D.H. Lawrence
The Homestead

About 1840, a canal was constructed across the meadows of the
Marsh Farm, connecting the newly-opened collieries of the Ere-
wash Valley. A high embankment travelled along the fields to carry
the canal, which passed close to the homestead, and, reaching the
road, went over a heavy bridge.

So the Marsh was shut off from Ilkeston, and enclosed in the
small valley bed, which ended in a bushy hill and the village spire
of Cossethay.

The Brangwens received a fair sum of money from this trespass
across their land. Then, a short time afterwards, a colliery was
sunk on the other side of the canal, and in a while the Midland
Railway came down the valley at the foot of the Ilkeston hill, and
the invasion was complete. The town grew rapidly, the Brangwens
were kept busy producing supplies, they became richer, they were
almost tradesmen.

Still the Marsh remained remote and original, on the old quiet
side of the canal embankment, in the sunny valley where slow
water wound along in company of stiff alders, and the road went
under ash-trees past the Brangwens' garden gate.

But, looking from the garden gate down the road to the right,
there, through the dark archway of the canal's square aqueduct,
was a colliery spinning away in the near distance, and further, red,
crude houses plastered on the valley in masses, and beyond all,
the dim smoking hill of the town.

The homestead was just on the safe side of civilization, outside the
gate. The house stood bare from the road, approached by a straight
garden path, along which at spring the daffodils were thick in green
and yellow. At the sides of the house were bushes of lilac and
guelder-rose and privet, entirely hiding the farm buildings behind.

At the back a confusion of sheds spread into the home-close from out of two or three indistinct yards. The duckpond lay beyond the furthest wall, littering its white feathers on the padded earthen banks, blowing its stray soiled feathers into the grass and the gorse bushes below the canal embankment, which rose like a high rampart near at hand, so that occasionally a man's figure passed in silhouette, or a man and a towing horse traversed the sky.

At first the Brangwens were astonished by all this commotion around them. The building of a canal across their land made them strangers in their own place, this raw bank of earth shutting them off disconcerted them. As they worked in the fields, from beyond the now familiar embankment came the rhythmic run of the winding engines, startling at first, but afterwards a narcotic to the brain. Then the shrill whistle of the trains re-echoed through the heart, with fearsome pleasure, announcing the far-off come near and imminent.

As they drove home from town, the farmers of the land met the blackened colliers trooping from the pit-mouth. As they gathered the harvest, the west wind brought a faint, sulphurous smell of pit-refuse burning. As they pulled the turnips in November, the sharp clink-clink-clink-clink-clink of empty trucks shunting on the line, vibrated in their hearts with the fact of other activity going on beyond them.

Idris Davies
High Summer on the Mountains

High summer on the mountains
And on the clover leas,
And on the local sidings,
And on the rhubarb leaves.

Brass bands in all the valleys
Blaring defiant tunes,
Crowds, acclaiming carnival,
Prize pigs and wooden spoons.

Dust on shabby hedgerows
Behind the colliery wall,
Dust on rail and girder
And tram and prop and all.

High summer on the slag heaps
And on polluted streams,
And old men in the morning
Telling the town their dreams.

George Owen of Henllys
This Black Labour

Next to the wood, or rather to be preferred before it for smell, is the coal fire for the generality of it, as that which serves most people and especially the chief towns. This coal may be numbered as one of the chief commodities of this county and is so necessary as without it the county would be in great distress. It is called stone coal for the hardness thereof and is burned in chimneys and grates of iron, and being once kindled gives a greater heat than light and delights to burn in dark places. It serves also for smiths to work, though not so well as the other kind of coal, called the running coal, for that, when it first kindles, it melts and runs as wax and grows into one clod, whereas this stone coal burns apart and never clings together. This kind of coal is not noisome for the smoke, nor nothing so loathsome for the smell as the running coal is, whose smoke annoys all things near it, as fine linen and men's hands that warm themselves by it, but this stone coal yields in a manner no smoke after it is kindled, and is so pure that fine cambric or lawn is usually dried by it without any strain or blemish, and is a most proved good drier of malt, therein passing wood, fern or straw. This coal for the rare quality thereof was carried out of this county to the city of London, to the late Lord Treasurer Burleigh, by a gentleman of experience to show how far the same excelled that of Newcastle, wherewith the city of London is served, and I think if the passing were not so tedious, there would be a great use made of it. And now that I am come to entreat of this our county's coal, I must remember my promise made before, where I spoke of the veins of limestone which I said was found to accompany the veins of coal and therefore I will in as few words as I can show you the natural course of this coal and how the same does accompany the limestone vein. I said before that I found out two veins of limestone to have their original here in Pembrokeshire, and that their course holds eastward, as before I have declared at large, between both which veins of limestone the coal is found to follow, but not so open as the limestone in every place with the limestone, but in many places where the stone shows, the coal hides itself, and where the coal is found, sometimes the limestone lurks underground, and in many places they are found near together. And first our coal has been found near Talbenny and so follows on to Johnston and there found; then at Freystrop, great store, and so at Picton. It is also found by the southern vein of limestone at Jeffreston, and from thence to Begelly. This first vein of coal follows the first vein of limestone, keeping on the south side of it to the water, and so to the mouth of the Towy over the bar of Carmarthen ....

This other vein of coal which I spoke of at Jeffreston accompanies the second vein of limestone on the north side thereof within half a mile of the limestone and passes east to Saundersfoot and there accompanying the limestone to the sea ....

The digging of this coal is of ancient time used in Pembrokeshire, but not in such exact and skilful sort as now it is, for in former time they used no engines for lifting up the coal out of the pit but made their entrance slope, so as the people carried the coal upon their backs along stairs which they called landways, whereas now they sink their pits down right four square, about six or seven foot square, and, with a windlass turned by four men, they draw up the coal, a barrel-full at once by a rope. This they call a downright door. The lords of the land have either rent, or the third barrel after all the charges of the work deducted. The coal is first found out by a small appearance thereof, which they call the edge, which being found they search which way the vein leans and on the contrary side they begin to sink, for the coal is found to lie slope in the ground, and seldom downright. The coal being found, the workmen follow the vein every way until it end or be letted by water or rock. The vein for the most part will not be passing five or six foot deep, so that the coal is carried stooping, for they commonly leave a foot of coal in the bottom undug to serve for a strong foundation, except they find the rock under foot, which they call the dunstone, which if they find then they dig clean all the coal, and further than that stone they look for no coal. And overhead they are driven to timber their working to keep the earth from falling, which is chargeable, but in some grounds they have a rock above and then they save much labour and cost in sparing of timber.

In these works the water springs are troublesome, which they avoid by sinking a great pit right under the door to which all the water will run and from thence draw it up with a windlass by barrels or else by making a level as they call it, which is by a way dug underground somewhat lower than the work to bring a passage for the water. This is very chargeable and may cost sometimes £20 and oftentimes more. They now most commonly sink down right twelve, sixteen or twenty fathoms before they come to the coal, whereas in old time four fathoms was counted a great labour. When they find it, they work sundry holes, one for every digger, some two, some three or four, as the number of diggers are, each man working by candlelight and sitting while he works. Then they have bearers, which are boys that bear the coal in fit baskets on their backs, going always stooping by reason of the lowness of the pit. Each bearer carries this basket six fathoms where, upon a bench of stone, he lays it where meets him another boy with an empty basket which he gives him and take that which is full of coal and carries it as far, where another meets him, and so till they come under the door where it is lifted up. In one pit there will be sixteen persons, whereof there will be three pickaxes digging, seven bearers, one filler, four winders, two riddlers, who riddle the coal when it is a-land, first to draw the small coal from the big by one kind of riddle, then the second riddling with a smaller riddle with which they draw small coal for the smiths from the culm, which is indeed but very dust, which serves for lime burning. These persons will land about eighty or hundred barrels of coal in a day. Their tools about this work are pickaxes with a round poll, wedges and sledges to batter the rocks that cross their work.

All times of the year is indifferent for working but the hot weather is worst by reason of sudden damps that happen, which oftentimes causes the workmen to swoon and will not suffer the candles to burn, but the flame, waxing blue of colour, will of themselves out. They work from six o'clock to six o'clock and rest an hour at noon and eat their allowances, as they term it, which is a halfpenny in bread to every man and four pence in drink among a dozen. This is of custom on the charge of the pit, although they work on their own charge. All their work is by candlelight throughout the year.

The coal they find is either an ore coal, a string or a slatch, as I have learned their terms. The ore is the best and is a great vein spreading every way and endures longest. The string is a small narrow vein sometimes two, three of four foot in bigness and runs downright and is always found between two rocks. A slatch they call a piece of coal by itself found in the earth and is quickly dug about and no more to be found of that piece.

The first of these three sorts is the best, then the next, and the last accounted worst of all.

The dangers in digging this coal is the falling of the earth and killing of the poor people, or stopping of the way forth, and so die by famine, or else the sudden irruption of standing waters in old works. The workmen of this black labour observe all abolished holy days and cannot be weaned from that folly.

John Ormond
My Grandfather and His Apple-Tree

Life sometimes held such sweetness for him
As to engender guilt. From the night vein he'd come,
From working in water wrestling the coal,
Up the pit slant. Every morning hit him
Like a journey of trams between the eyes;
A wild and drinking farmboy sobered by love
Of a miller's daughter and a whitewashed cottage
Suddenly to pay rent for. So he'd left the farm
For dark under the fields six days a week
With mandrel and shovel and different stalls.
All light was beckoning. Soon his hands
Untangled a brown garden into neat greens.

There was an apple-tree he limed, made sturdy;
The fruit was sweet and crisp upon the tongue
Until it budded temptation in his mouth.
Now he had given up whistling on Sundays
Attended prayer-meetings, added a concordance
To his wedding Bible and ten children
To the village population. He nudged the line,
Clean-pinafored and collared, glazed with soap,
Every seventh day of rest in Ebenezer
Shaved on a Saturday night to escape the devil.

The sweetness of the apples worried him
He took a branch of cooker from a neighbour
When he became a deacon, wanting
The best of both his worlds. Clay from the colliery
He thumbed about the bole one afternoon
Grafting the sour to sweetness, bound up
The bleeding white of junction with broad strips
Of working flannel-shirt and belly-bands

To join the two in union. For a time
After the wound healed the sweetness held,
The balance tilted towards an old delight.

But in the time that I remember him
(His wife had long since died, I never saw her)
The sour half took over. Every single apple
Grew — across twenty Augusts — bitter as wormwood.
He'd sit under the box-tree, his pink gums
(Between the white moustache and goatee beard)
Grinding thin slices that his jack-knife cut,
Sucking for sweetness vainly. It had gone,
Gone. I heard him mutter
Quiet Welsh oaths as he spat the gall-juice
Into the seeding onion-bed, watched him toss
The big core into the spreading nettles.

Dennis Potter
The Small Mines of the Forest

Sitting in 'The Globe', on the other side of my father, was Teddy
Gwilliam, a man who holds himself like a young officer in the
Guards, for all that he has a large family and, unlike such young
gentlemen, works on his knees in one of the small, licensed pits
which are another characteristic of the Forest of Dean, still
thriving like tiny piglets round the great black half-dead sows of
the bigger mines. He works it with his two brothers, and they live
by what coal can be got out, which sometimes, when picking
through earth and rock, or when a promising seam peters out, can
be nothing at all for days or even weeks on end, but at other times
is good enough or promising enough to allow a longer break for
'bread' and more of a chat in the little wooden hut they have
erected near the black cavity in the earth which slants into the
slope of a tree-covered hill.

Teddy has a natural, impressive dignity which deserves better
than my comparison with a Guards officer, for his bearing is quiet,
helped no doubt by a high, well-shaped forehead from which his
hair has retreated. He talks rather in a monotone, but one is rarely
likely to be bored by it because he is one of those rare individuals
for whom almost everything has some sudden and special interest,
some unusual angle. He might almost have come from a George

Eliot novel: most of all, he is continually fascinated by the skills and the peculiarities of his job, which he appears to regard as a privilege rather than a burden — to be cussed at, of course, and be disappointed in too many times, but all the same a privilege and a freedom, a regular and complete means of expression. As a consequence, his attitude to his work, the attitude of a proud and almost self-sufficient craftsman, is anachronistic to modern ears. (The sense in which it could be said 'We are all workers now' is only possible if we are all shysters now.)

The small mines of the Forest, employing two, four or even a dozen men, have been part of the local landscape for centuries, racking houses, killing workers and depositing little mole-like tumps with extravagant frequency. They would really need a book on their own to be understood properly, existing by a combination of 'ancient privilege' and the type of land and land-ownership found in the area. Their names have a strange ring about them, like New Fancy or Ready Penny, and in the past but much less so today, they were important in local life, rarely producing great wealth, yet emphasizing the values of independence and village identity. (It will not escape the reader's attention, I suppose, that I have difficulty in describing or explaining these qualities, for you cannot build a jig-saw by describing its pieces.)

These small pits are subject to inspection and national safety regulations, but in other ways remain 'private' — very much at the small workshop stage of the industrial revolution, rather as the ownership of land in the Forest, which belongs mainly to the Crown, remains at an even earlier stage. The pit worked by the Gwilliam brothers — one tall and dignified, one stocky and ebullient, a good rugby player, and the other shy and impressively strong — is at Braceland, half a mile from Berry Hill and Ninewells. It has the cinematic appearance of a small prospector's camp in the Rocky Mountains, for the working area is swallowed up in the trees, some of which have sagged wearily towards each other as a result of the steady eating away of rock and soil down beneath their roots. The land is unsteady here, with small gradients working against the slope, and a long stony fall to the Wye a mile to the west. Two tub-wide tunnels are driven into the bank, ultimately connecting with each other as they creep into the darkness: there is, therefore, an escape route and a freer flow of air. A small thread of railway emerges from each to converge again on a pile of damp, messy-looking rubble which turns out to be small coal. There was mud everywhere, of a rich, oozy, squelchy kind. Props of timber and neat stacks of chunky logs lay in the foreground, an axe was stuck

into a bigger log, and a cloud of almost pure white wood smoke
hung in the drooping branches of the tree which acts as a second
roof above the small timber shack the brothers use as office,
planning centre, eating place and cloakroom.

In the shack, which is placed on one of the few bits of level
ground, a platform above the valley, they keep a small trunk which
has a carefully rolled, pen-drawn map, neatly shaded to show what
coal they might expect to find on this site. From the map, it is
clear that the area has been worked before, and Ted told me he
was going back to what his father had left years ago.

'I'm after what fayther left long before the war.'

It was in winter when I first scraped up the tunnel to the place
where the brothers were grunting with each chink of their tools,
biting into the dark vein which seeps out of the earth and rock like
a slow, faint stain. I would not have recognized it as coal, imprisoned in the sides of the face. Ted shadowed up before me like a
bent wizard in a magic lantern show, and there was a faint trickle
of water.

'Nervous?' Ted asked, certain in his mind that I was. I was.

'No,' I said, for reasons of pride, 'it's rather cosy up in here, en it?'

My back was already wet, and I was aching with the effort of
keeping bent. I was also once kicked rather savagely in the knee
when playing rugby for Berry Hill against another Forest of Dean
side — an entirely predictable injury, but not less painful — and
it hurts when I keep it bent for very long. After twenty minutes,
let alone a full day every day, I was an old crock, and soon wanted
to creep back to the opening and the light. But with a genuinely
captive listener, Teddy kept talking, the steady quiet rhythm of
his speech mounting every now and then to enthusiasm as he
described what he was about, or softening into a slightly amused
concern as I grunted and shifted uncomfortably, holding on to the
clammy, cold wall of the tunnel.

'Look,' my captor was explaining, a strange note in his voice,
'you can see here where the old men have been.' He pointed to
some rough, barely discernible lines in the rock to one side of him,
his body stretching forward to the limits of stability while still
balanced on the one knee. The marks looked as though they had
been made by a frightened cat trying to claw its way back to safety.

'What old men?'

'Oh, years and years back, o'but. Most of this land here has been
gone over and gone over till there yunt all that much coal for we
poor lot to get out, I can tell tha.' Teddy laughed with admiration
for the old Foresters, and with more wonder than regret. ''Sknow

what? I should like to go back, just once, just for a bit, and see
how they managed. They must have been marvellous people,
Den.' He shifted. 'Oy, they must have bin. They didn't have the
same tools as we had, mind — practically no metal. We found an
old wooden spade down here — oy, it makes you wonder, it do.'
He was silent again.

Soon, they were back at work, the three of them, talking finished
for a while. Each of the brothers specialized in a different aspect
of the job, one feeding the coal back to the other in a steady,
crumbly mount, and the third bringing in the timber or pulling
out the fat, heavy iron tubs, trundling with a slow clatter along the
small line curving out from the tunnel. Their cap lamps glowed
and fluttered in the dark, approached and retreated, blacked out
and then flared up again, and the tunnel edged its way pickshaft
lengths deeper into the hill, further below the tree roots and the
rotting layers of dead winter leaves.

Afterwards, in the hut, the early gloom already overtaking and
filling in the surrounding woods, showing up boldly the yellow
square of light from the only house that could be seen, a few
hundred yards below the present levels, the brothers talked about
their work and their leisure. They were very impressive, sure of
what they were doing and why they were doing it. I cannot think
that there will be many others like them from my generation in
the Forest of Dean.

'No, I couldn't work in a factory,' Teddy was as emphatic as it
is possible to be, 'I couldn't abear it. Clocking in and clocking out.
Never in control of what you'd want to do, always at somebody
or other's beck and call.' Harold and Dennis, his brothers, agreed,
and explained the virtues of being craftsmen working to their own
pace and their own needs, cruelly hard on some days when a piece
of ground was difficult, but more relaxed on others when the coal
was coming out in easier fashion. 'Nobody publishes our absentee
figures, or sends away a little card to some big office.' They were
acutely conscious of the silly hierarchies they would have to
encounter in the local factories, and intended to avoid the men-
tality behind the terminology of 'staff' and 'workers' so disliked
by the miners from the big pits when they had changed their jobs.
They didn't mind the hard or dirty work of a factory, if that is
what is was, but they have rejected the things which went with it.

Ted was not greatly impressed with many of the changes which
have made their onslaught on the Forest of Dean, precisely
because they involved some challenge to his individuality and his
dignity. Let the coloured balloons float down on the ballroom

dancers, he said, 'We shan't have a works dinner!' His life, his work and his values, like those of his brothers, remained unchanged, convincing himself (and me) that he had managed to hold on to some quality the rest might have let slip.

As elsewhere, of course, there is an uneasy relationship between the new and the stubborn but virtually doomed habits of the old. Unfortunately, we tend to see only the perversity or the quaintness of the latter, neglecting everything but the obvious incongruity. The Gwilliam brothers, and some few others of their kind in the Forest of Dean, act and think as though their labour was the most essential part of their personality, owed to themselves and not marketable by the demands of others, using it as a defence as well as an expression. 'Independence' is a popular claim in the Forest, but it is a rather meaningless concept when applied to the life of a working man, or to almost any of us; Teddy Gwilliam used the word, in a valid but retreating sense, to mean freedom from the irksome evasions of the capital-labour relationship which is inevitably seen in the life and spirit of the local factories; freedom from their hierarchies and their power. Perhaps it is too much to ask of the world, but meanwhile, there is the damp, curving tunnel and the mounds of small coal, the work of a lifetime.

Barry Hines
Royal Visit

In the pit bottom, a miner was fitting a little pair of velvet curtains to the wall. He was adjusting the pelmet when the cage reached the bottom of the shaft and the onsetter let the men out. They walked as far as the curtain hanger then stopped to see what he was doing.

Alan said, 'What's this then, the drapery department?'

Ronnie tried to open one of the curtains but the hanger pushed his hand away and rearranged the fold.

'Wait a minute. It's not finished yet.'

'Hurry up then. We want to have a look.'

The hanger fiddled under the pelmet, then, with all the men looking on, pulled a cord and unveiled a stone plaque. It said:

TO COMMEMORATE
THE VISIT OF
HRH PRINCE CHARLES
TO MILTON COLLIERY
JUBILEE YEAR: JUNE 21ST 1977

The men applauded and cheered and called for more. Pleased with the smooth traction of the hooks on the rail, the hanger pulled the curtains shut for another go.

A Deputy, walking towards the pit bottom at the end of his shift, saw them standing there, saw them applauding and laughing, and when he reached them, he looked between their heads to see what they were looking at.

'What you lot gawping at?'

Harry turned at the voice at his shoulder.

'Hey up, Ken. You're just in time, we're waiting for the Punch and Judy to start.'

'I'll give you Punch and Judy ...' The curtain hanger did not reveal the plaque again now that the Deputy had arrived. 'And listen. I hope you men have seen that notice in the baths about swearing. Remember them tannoys are never switched off. We don't want a mouthful of foul language filling the pit all of a sudden while He's being shown round.'

'What about, flipping and bum? Are they swearing?'

The Deputy pointed his stick at Alan, almost touching his chest with it.

'You don't be cheeky, lad, or I'll send you out of the pit. You know what I mean. I'm talking about effing and blinding, that kind of thing.'

It could have been a teacher admonishing a pupil. But Alan just grinned at him: it took more than a bit of cheek to an official to get suspended these days.

As the men moved off down the roadway, Syd turned back to the Deputy who was now enjoying a private viewing of the plaque.

'You can rely on us, Ken. Don't you worry about that. I mean, we wouldn't want Him going home thinking that anybody used bad language down here, would we?'

The deputy waved Syd on his way with his stick.

'Go on, bugger off! And make sure you get some bloody work done for a change!'

Near the pit bottom, the roadway was as wide and high as a tube station. The white brick walls had recently been re-painted and the tunnel was well lit by neon strips on the roof. The men passed rows of loaded pit tubs waiting to go to the surface, then passed through an air door, each man holding it open for the next because of the strong draught. The pit was ventilated by a series of these doors. They were always kept shut, to stop and re-direct the flow of air so that it was forced into all the workings and did not blow the shortest route along the main roadways, and straight back to the surface up the upcast shaft.

The last man through let the door bang shut behind him. The men stopped at this side of the door and waited for the paddy train to arrive to take them closer to their work places, which were more than two miles away from the pit bottom.

While he was waiting, Syd walked up the roadway a few yards to watch a miner who was painting the flanges of an airpipe fastened along one wall. He was concentrating on the job and taking great care not to touch the pipe itself which had been painted a different colour. Syd stood quietly behind him and watched, not daring to speak in case he made the painter smudge his work.

When he had finished, the painter stepped back to consider the effect.

Syd nodded appreciatively. 'Very nice, Maurice. They're bringing two tubs of flowered wallpaper down on the next draw. The Deputy says you've to start putting it up as soon as you've finished here.'

The painter stirred his paint and prepared to move or to the next flange.

'I'll tell you what. I'd sooner put paper up, than rings up like you lot any day.'

Then they heard the paddy train coming. Two lights, one above the other, approached them out of the darkness, and as they came closer, they could see that the top light was the driver's cap lamp, and the bottom one was attached to the front of the train. The paddy was rope hauled, and had open carriages with facing seats like a miniature railway train in a pleasure garden.

The train stopped; the driver got down, unclipped the lamp from the front, then walked the length of the train and clipped it to the other end. It was now ready to go back the way it had come. The driver took his seat and the men took theirs. Harry asked where the Royal Visitor would be sitting. The driver showed him, and Harry sat down in His place. He shuffled around in mock discomfort then knocked on the wooden board disapprovingly.

'I hope you're getting Him summat softer than these boards to sit on, Eddie.'

'They've had some foam rubber cushions made, and covered them in material.'

'I'm glad to hear it. We don't want Him sitting on owt cold and getting piles.'

Ronnie took the seat next to Harry, directly behind the driver.

'Somebody told me that they were converting an old tub, and they were going to find four white pit ponies to pull it.'

Syd asked the flange-artist if he could borrow his paint and brush, then walked round the front of the paddy train and started to paint something across the front of it. Alarmed, the driver stood up and leaned over the front to try to read what Syd was painting.

'Hey up, Syd. You'll be getting me shot.'

'What's up with you? They'll be painting it before the Day, won't they?'

'I know, but that's not the point. The Manager'll play bloody hell with me if he sees it.'

'He can't blame you, Eddie. You were having your snap somewhere, weren't you? You never saw a thing.'

Alan and Albert climbed down off the paddy and walked round the front to see what Syd was writing. Harry asked them, but they waited until Syd had finished before they told him.

Then Albert said, 'Royal Coach.'

Syd returned the brush and paint and three men got on the train. They were ready to go now, and the driver gave the starting signal by reaching up with a metal rod and pushing together two electrified wires which ran close to the roof above the paddy track. He released the brake and the train jerked forward with sufficient force to tilt some of the men's helmets back. They cursed him, readjusted them, and the train gathered speed. Every time it passed a group of miners at the side of the track, Harry gave them a fixed smile and a regal wave. Syd focused his lamp on the airpipe fixed to the wall and counted all the blue flanges which had been painted along it.

The paddy stopped several times to let men out, and at one of these stopping places the driver pointed to a rough cross painted on the roof above him.

'See that? I've got to stop straight underneath that cross on the Day.'

The others did not know what he was talking about.

'Well, on the Big Day, when the official party visits the face, this is where they get off. That cross is dead opposite the gate end. I've to practise stopping the front coach straight underneath it, so that when He gets off, He can walk straight down to the face.'

The ill-lit junction and the dirt on the driver's face, gave enough cover to keep the others guessing whether he was serious or not.

'It's right. Forbes and Carter came down yesterday and worked it out. What a pantomime. You've never seen anything like it. Carter was on the blower giving instructions to Ned Butler up at the haulage engine, and I was driving:

"Back a bit Ned!" he kept shouting. "That's it! That's it! Now stop!... Whoa! Whoa! Too far! You've gone too far! Bloody hell, Ned, what you playing at?... Now forward.... A yard.... That's it.... A bit further.... Right. There. That's it.... That's it, Ned! Whoa! Whoa, for Christ's sake, you useless pillock!... Ned, do you call that a yard? I wish I'd a yard of baccy as long, that's all. Now back a bit, and let's get it right this time. And I'll tell you what, if you don't get it right this time you'll not be on that bloody job on the Day. I'll have you cleaning the shit houses out instead! You might be a bit more useful in there!"

'We must have been at it half an hour before we got it right, and Forbes was sat here at the side of me all the time. He was having a fit. The air was blue I can tell you.'

Harry, still sitting in the royal place, shone his lamp up at the cross. A plumb bob dropped from it would have hit him.

'Would you believe it? They don't mean Him straining Himself do they?'

He leaned over the side and looked down at the twelve-inch drop from the carriage floor to the ground.

'It's a wonder they haven't had the joiners making a set of steps for Him to step down off.'

The driver rejected that notion with an emphatic shake of the head.

'Oh no, I've heard they sent a letter from the Palace saying that they don't want any special privileges like that.'

Harry nudged Ronnie, then got off the paddy.

'Big hearted them Royals, Ronnie. There's no doubt about that.'

Syd and the other rippers stayed on the paddy until it reached the end of the track. They made the usual jokes about stopping on and going straight back, before climbing down and starting the half-mile walk to the ripping edge. The only light they had now was from their cap lamps and they walked in single file between the wall and the conveyor belt, concentrating on their footing because of the uneven ground. This concentration made them quiet; and the silence soon made them miserable, because it gave them a chance to think of the shift ahead, and how long it would seem before they were walking back along the road at the end of it.

Norman Nicholson
Cleator Moor

From one shaft at Cleator Moor
They mined for coal and iron ore.
This harvest below ground could show
Black and red currants on one tree.

In furnaces they burnt the coal,
The ore was smelted into steel,
And railway lines from end to end
Corseted the bulging land.

Pylons sprouted on the fells,
Stakes were driven in like nails,
And the ploughed fields of Devonshire
Were sliced with the steel of Cleator Moor.

The land waxed fat and greedy too,
It would not share the fruits it grew,
And coal and ore, as sloe and plum,
Lay black and red for jamming time.

The pylons rusted on the fells,
The gutters leaked beside the walls,
And women searched the ebb-tide tracks
For knobs of coal or broken sticks.

But now the pits are wick with men,
Digging as dogs dig for a bone:
For food and life *we* dig the earth —
In Cleator Moor they dig for death.

Every wagon of cold coal
Is fire to drive a turbine wheel;
Every knuckle of soft ore
A bullet in a soldier's ear.

The miner at the rockface stands,
With his segged and bleeding hands
Heaps on his head the fiery coal,
And feels the iron in his soul.

Joe Corrie
A Cageload of Men

Just like a truck load of cattle,
    Sixteen crushed on at a time,
The yawning abyss beneath them,
    Awaiting the 'bottomer's' chime,
To leave all the glories of nature,
    And toil in the muck and the grime.

Hard-handed stalwarts of labour,
    Nurtured to grin and to bear,
Seldom a thought of the danger
    That haunts every corner down there,
Praying to Christ it was 'lowsing',
    But not in the language of prayer.

Nipper so proud to be working,
    Grandad with hair like the snow,
One with his eyes on the heavens,
    One with his eyes on below,
Free to stay up if they wish it,
    But hunger, ah! both of them know.

One with the cares of a household,
    Weary and sick of it all,
The best of his years he has given,
    And now with his back to the wall.
Haunted with fears of the future,
    Dreading how far he will fall.

Clang! goes the 'bottomer's' signal,
    Down, strangely silent, they go,
In comes another mixed cageload,
    Each with a number to show,
Cogs in the wheel of Corruption,
    Grinding so sure and so slow.

John L. Hughes
Nothing Special

There being nothing special much around this town. Not like London nor Paris nor Rome nor Lisbon nor Washington. Not like Cardiff nor Tenby nor Bangor nor plenty of towns you can see anywhere.

No one particular item making you sit up sudden like you never believed what you saw first time off. Not a single special thing for making passing (on to God knows where) strangers remember where they been. No special beauty. No special ugliness. No sudden visions. And no sudden blindness as ever they (them passing strangers) could recall.

Nothing special much around this town for filling empty exiled nights and minds with nostalgia nor things hard to forget a hundred a thousand a hundred thousand miles away. Nothing memorable you understand. No one peculiar thing them passing eyes could retain for ever. No wide gasping geographical statement marked deep across the landscape making them (them passing strangers) whisper all holy:

Only God could have done that.

Now tell me there is no God.

God been poking his fingers up this place.

God definitely done that mountain (or that valley or that cliff or that forest or that sky).

Even the name of this place is forgettable. Pontypridd. A shambles of mystic Welshness. Pontypridd. Something to do with a bridge (there is a bridge). Pontypridd. Something to do with the earth (black stuff). Who could remember such a name? Who could care?

There being nothing special much around this town. Nothing at all. Except perhaps the river. Maybe just that.

The River Taff. Swilling down from Merthyr same as some kind of whip. Dirty candle-coloured by day down through Aberdare Mountain Ash Abercynon and Cilfynydd in a torrent. In a welter of torrents. Grunting sucking lashing whirlpools blackened through by mining trash and coal no man could burn.

Twisting fast down deep inside the guts of Pontypridd. Towards Cardiff. Carrying things. Always carrying things.

Dead terrier. Four headless chickens. Five loaves (from Wonderloaf) and two fish (from Plowman's) floating in circles amid surface scum just off that Marks and Spencer back wall. Two mangled oil cans from Texaco (by way of Halfords) and one sodden *Western Mail* hanging same as nests on the pea-shooter beds fringed haphazard around Ynysangharad Park. And a bucket. And a bugle. And a birthday card. And a trouser leg with the turn-up down. And a boy and his ferret hunting for rats.

Found a bike down there once.

Only needed a chain.

Found a chain down there once.

Only needed a bike.

And in flood that river is mighty. A frightening thing. Unstoppable (if ever man had a mind). Creaking heavy against them quaking banks where houses stand. And people stand. Staring. Up at the rain. Down at the river. Across at each other. Wondering. With tomboys chucking stones snick snick into the grey flow. And today it is raining as last night it rained and the day before. And the night before that day. For this town is Pontypridd. Where they know all about rain.

And it was raining when you got born. And it will be raining when you die. If you die in Pontypridd.

And mister I can tell you for nothing as how when that day comes you will definitely have enough on your plate working your dead mouth on a fiddle into Heaven never mind that hissing rain.

With thoughts like that definitely giving you the willies now and again seeing as how you are not ready for no bloody drop this fair side of three score years and ten. With a stack of functioning left for doing inside your skull not to mention all through your medium rare skin. Leaving them passing strangers to keep on passing by this forgettable place in a gust of wind and rain. On towards London or Paris or Rome or Lisbon or Washington or Cardiff or Tenby or Bangor or anywhere else. For this town is Pontypridd where the gods brought your soul at the beginning of time. And nailed it to a wall.

Walking on sharp down through Taff Street in a December drizzle towards your wife. Towards Woolworth's where Rachel been waiting an hour.

Hello mister high and bloody mighty.

Where the hell you been?

Got held up isn't it.

Pay was late coming.

You smell of drink.

I been here an hour.

Dying for the toilet.

You got some money?

Yes.
For Christ sake Ben.
Giving her forty pounds in your brown envelope hall-marked
and stamped with the compliments of the National Coal Board
minus the docket away up inside your rattling hat. And watching
her run for the Cilfynydd bus jostling through that steaming
Christmas crowd like they are litter.

Dannie Abse
Welsh Valley Cinema, 1930s

In The Palace of the slums,
from the Saturday night pit,
from an unseen shaft of darkness
I remember it: how, first, a sound
took wing grandly; then the thrill
of a fairground sight — it rose,
lordly stout thing, boasting
a carnival of gaudy-bright,
changing colours while wheezing out
swelling rhonci of musical asthma.

I hear it still, played with panache
by renowned gent, Cathedral Jones,
'When the Broadway Baby Says Goodnight
it's Early in the Morning' — then he and it
sank to disappear, a dream underground.

Later, those, downstairs, gobbing silicosis
(shoeless feet on the mecca carpet),
observed a miracle — the girl next door,
a poor ragged Goldilocks,
dab away her glycerine tears
to kiss cuff-linked Cary Grant
under an elegance of chandeliers.
(No flies on Cary. No holes in *his* socks.)

And still the Woodbine smoke swirled on
in the opium beam of the operator's box
till THE END — of course, upbeat.
Then from The Palace, the damned Fall,

the glum, too silent trooping out
into the trauma of paradox:
the familiar malice of the dreary,
unemployed, gas-lamped street
and the striking of the small Town's clocks.

Idris Davies
Send Out Your Homing Pigeons, Dai

Send out your homing pigeons, Dai,
Your blue-grey pigeons, hard as nails,
Send them with messages tied to their wings,
Words of your anger, words of your love.
Send them to Dover, to Glasgow, to Cork,
Send them to the wharves of Hull and of Belfast,
To the harbours of Liverpool and Dublin and Leith,
Send them to the islands and out of the oceans,
To the wild wet islands of the northen sea
Where little grey women go out in heavy shawls
At the hour of dusk to gaze on the merciless waters,
And send them to the decorated islands of the south
Where the mineowner and his tall stiff lady
Walk round and round the rose-pink hotel, day after
    day after day.
Send out your pigeons, Dai, send them out
With words of your anger and your love and your
    pride,
With stern little sentences wrought in your heart,
Send out your pigeons, flashing and dazzling towards
    the sun.

Go out, pigeons bach, and do what Dai tells you.

You Will Get Your Chain of Gold, My Lad: Starting Out

Joseph Skipsey
Mother wept, and father sighed

Mother wept, and father sighed;
  With delight a-glow
Cried the lad, 'To-morrow,' cried,
  'To pit I go.'

Up and down the place he sped, —
  Greeted old and young,
Far and wide the tidings spread, —
  Clapt his hands and sung.

Came his cronies some to gaze
  Wrapt in wonder; some
Free with counsel; some with praise;
  Some with envy dumb.

'May he,' many a gossip cried,
  'Be from peril kept;'
Father hid his face and sighed,
  Mother turned and wept.

Barry Hines
The Interview

'Now then, Casper, what kind of job had you in mind?'
  He shunted the record cards to one side, and replaced them with a blank form, lined and sectioned for the relevant information. CASPER, WILLIAM, in red on the top line. He copied age, address and other details from the record card, then changed pens and looked up.
  'Well?'
  'I don't know, I haven't thought about it right.'
  'Well you should be thinking about it. You want to start off on the right foot, don't you?'
  'I suppose so.'
  'You haven't looked round for anything yet then?'
  'No, not yet.'
  'Well what would you like to do? What are you good at?'
  He consulted Billy's record card again.

'Offices held .... Aptitudes and Abilities ... right then ... would you like to work in an office? Or would you prefer manual work?'

'What's that, manual work?'

'It means working with your hands, for example, building, farming, engineering. Jobs like that, as opposed to pen pushing jobs.'

'I'd be all right working in an office, wouldn't I? I've a job to read and write.'

The Employment Officer printed MANUAL on the form, then raised his pen hand as though he was going to print it again on the top of his head. He scratched it instead, and the nails left white scratches on the skin. He smoothed his fingers carefully across the plot of hair, then looked up. Billy was staring straight past him out of the window.

'Have you thought about entering a trade as an apprentice? You know, as an electrician, or a bricklayer or something like that. Of course the money isn't too good while you're serving your apprenticeship. You may find that lads of your own age who take dead end jobs will be earning far more than you; but in those jobs there's no satisfaction or security, and if you do stick it out you'll find it well worth your while. And whatever happens, at least you'll always have a trade at your finger tips won't you?...

'Well, what do you think about it? And as you've already said you feel better working with your hands, perhaps this would be your best bet. Of course this would mean attending Technical College and studying for various examinations, but nowadays most employers encourage their lads to take advantage of these facilities, and allow them time off to attend, usually one day a week. On the other hand, if your firm wouldn't allow you time off in the day, and you were still keen to study, then you'd have to attend classes in your own time. Some lads do it. Some do it for years, two and three nights a week from leaving school, right up to their middle twenties, when some of them take their Higher National and even degrees.

'But you've got to if you want to get on in life. And they'll all tell you that it's worth it in the end .... Had you considered continuing your education in any form after leaving? ... I say, are you listening, lad?'

'Yes.'

'You don't look as though you are to me. I haven't got all day you know, I've other lads to see before four o'clock.'

He looked down at Billy's form again.

'Now then, where were we? O, yes. Well if nothing I've mentioned already appeals to you, and if you can stand a hard day's graft and you don't mind getting dirty, then there are good opportunities in mining....'

'I'm not goin' down t'pit.'

'Conditions have improved tremendously....'

'I wouldn't be seen dead down t'pit.'

'Well what do you want to do then? There doesn't seem to be a job in England to suit you.'

He scrutinized Billy's record card again as though there might be a hint of one there.

'What about hobbies? What hobbies have you got? Do you like gardening, or constructing Meccano sets, or anything like that?'

Billy shook his head slowly.

Anon.
Collier's Song

In the depth of a coal-pit a young lad grew ,
    The pride of his mother's care,
All round his waist the coal waggon belt he drew,
    And his shoulders were stripped quite bare;
Thus he toiled the day long as he crept through the bay,
    And chained to his load like an ass,
Though hard was his work yet too little was his pay,
    In youthful days to pass.
        Then how long will the colliers and waggoners stand
        The real black slaves o' their native land?

How many are found either buried or drowned,
    And all the whole are oft away swept,
Or they perish by damp or the fire of a train,
    Which brings to an awful death.
Then the mother may weep for her only son,
    And the wife for her husband dear.
Whilst our families starve in a pauperish home,
    And how little do our masters care.
        Then how long will the colliers and waggoners stand
        The real black slaves o' their native land?

The banksman at top he told us he knew,
    That union made us strong,
That each man to another should be true,

And I echo these words in song.
Then let us unite to seek for our rights,
And like Britons go hand in hand,
We deserve to be paid for the labour we have made
In the bowels of our own native land.
        Then how long will the colliers and waggoners stand
        The real black slaves o' their native land?

At five in the morn we leave our home,
Our wives and children dear,
And perhaps we may never return,
To our native homes so dear.
But all the long day we are deprived of light,
And of the warmth of the sun.
And whilst we are in rags in poverty we run,
And this is our native land.

Harold Heslop
Harton Colliery

By the New Year our family had reached crisis point and was in danger of imminent collapse. It had long been apparent that the lady my father had taken to wife had arrived at the conclusion that she had made a profound mistake. The life she had visualised when she stood by his side before the altar had not turned out at all well. Her ukase was published at a family gathering. I was her chosen victim. If I did not depart from under the parental roof, she would forthwith 'take her 'ook'.

I had no alternative, and so I packed my belongings into a couple of brown paper parcels and departed, leaving my elder brother to join the army. The eldest of the three youngest sons was soon to go, leaving my father to enjoy the luxury of keeping up appearances with a woman who luxuriated an ungovernable temper.

I found a job at Harton Colliery, Tyne Dock, South Shields.

At the age of sixteen I was on my own in a strange town. Luckily I found lodgings with a good proletarian family. I still consider that I was fortunate to be granted a lifelong place at the fireside of Billie and Emily Gibson, at 23 Marsden Street, Westoe, South Shields. Billie was a small-boned, ginger-headed man who hewed coal at Marsden Colliery. Emily was a fine, intelligent and coura-geous woman. I think that they took me in as a replacement for

one of their two eldest sons who had died some time just before
the war commenced. Under their roof I crouched among the
turmoils of mine and adolescence, alone, stupendously bereaved
of my family, without friends, relatives and confidants.

I still recall vividly my introduction to Harton Colliery.

The Harton Coal Company Limited was a vast mining operation carried out in the rich seams of excellent coal lying under the
county borough and the rural district of Shields. The tremendous
seams were mined at four separate pits, St Hilda, Harton, Marsden and Boldon. Each pit was geared to a daily production of three
thousand tons of coal. It was at that time the most majestic
complex in the county of Durham. Each pit was powered by
electricity throughout. The economic basis of the structure was a
secure and satisfactory coalfield. All the produced coal was screened
before being loaded into the company's own waggons, carried on
its own electrified railway to its own staithes near the mouth of
the Tyne, where it was loaded by its own coal trimmers into
colliers, big and small. The company's own railway serviced three
of the collieries, and the fourth, Boldon, had to obtain the services
of the then North Eastern Railway. The entire county borough
was conscribed, transport-wise, by the colliery railway.

Not one chimney honoured the company with a belch of smoke.
Each colliery was drawn by an impressive winding engine and an
enormous headgear that imposed its ever revolving wheels over a
scene of incessant labour activity. Even the subsidiary appurte-
nances complemental to the task of creating the coal commodity
were impressive. The pit buzzer was an electrified screech. Coal
drawing at each pit commenced at precisely six o'clock in the
morning and continued without pause until nine o'clock in the
evening, five days one week and an extra few hours on the
Saturday of next. During the year the pit made holiday on Xmas
Day, New Year's Day, Good Friday Easter Monday, Whit Monday
and August Bank Holiday Monday, and on no other day of the
year, except, of course the alternative Saturday morning. No
miner ever took a holiday; none had the money for such an
easement.

Three seams were being worked during the time I was there, the
Yard Seam, the Bensham Seam and the Hutton Seam. The shaft,
which was sunk to the middle seam, the Bensham, was two
hundred and forty fathoms deep. The coal won in the Yard Seam
was dropped to the Bensham, and that won in the Hutton Seam
was drawn up through a long drift. All the coal was a shining,
beautiful mineral.

It was on a Sunday night round about half past ten when I set off from my new home to work my first shift in the colliery. I made my way rather tentatively to the lamp cabin where I handed over my permission to take a safety lamp. It was a huge, tiled place holding well over two thousand safety lamps, all of one pattern, all made to burn oil. I received my lamp at the counter that ran the length of the room, and followed the men and boys into what proved to be a huge waiting room. At the doorway stood two men whose duties it was to test the lamps. This was done by thrusting the lamp into a chamber filled with gas from the local gas supply. The lamp was merely extinguished. At the other side of the room a youth manipulated an electrical gadget which relighted the lamp by creating an inner flame at the end of a piece of wire near the wick of the lamp.

After I had received my lamp from the youth another young fellow approached me and told me to stick by him. This I was most content to do, for I was feeling strangely perturbed by the immense crowd of men and youths. I was conducted through a set of revolving doors, up a flight of steps, through another door and on to the pit head. The steel headgear ran up from some place hidden below and out to the wheels, which were still enclosed within the huge building. A buzzer tore a great leaf out of the closing day and the ropes began to move. Over all there was the continuous roaring of the suction fan which was located some distance below the pit head. All this noise overpowered any which the cages and ropes might make. They seemed to slip silently upwards to receive their human freight.

Each cage carried two decks, and each deck accommodated twenty people. I accompanied my mentor into the bottom deck of the cage. The descent was quite slow. We passed into the great wind created by the suction fan, and at last we were stayed in our progress by a slight bump on to the timbers over the sump. The gates were opened and we all filed out into an arched cavern that was flooded with electric light. Here was space in which one could move. I was totally impressed, for this was something I had not expected. This was mining on an immense scale, novel and to me incomprehensible.

The air moving against me was quite warm. It flowed over the clad body like tepid liquid.

I kept close behind my new companion as we passed under the arched cavern. We passed through a series of wooden doors whose purpose was to isolate the intake from the return. To have had all six doors open at the same time would have caused a ventilatory catastrophe. We emerged upon a section of the shaft landing. Here there were more coal tubs standing on the one side fully laden and on the other side empty than I had ever seen in my life. I calculated that in a mine of the size of this one there must have been some thousands of such springless boxes. Each carried exactly ten hundredweight of coal. There was no noise. A gale of cold, fresh wind caught my breath as I slipped into the passage between the rows of tubs. We were not a great distance from the bottom of the downcast shaft where the coal was drawn. Here again I noticed that the bricklayer and the mason had been busy years before I intruded upon the scene. The tunnel was magnificently arched and safe. Its entire length was illuminated by electricity.

We walked against the wind. Everything about me was built on the large scale, and was set for rapid, unimpeded transportation. There was a neatness of organisation, a simple efficiency, about the place that fascinated me, almost shocked me. There was a total absence of squalor. There was quality as well as quantity embodied in this underground scene, tending towards the intensification of a process that could be felt although everything was still. Later I became used to this shaft bottom and able to appreciate the orderliness of the work. There were two levels of approach to the shaft, the upper one into which the loaded streams of tubs discharged themselves, and the lower into which the empty tubs were received from the cages. There were four decks to each cage, which carried two tubs to each deck. The act at the bottom was an almost automatic one of inserting the two tubs from each deck and expelling the two empty ones in one movement. The act at the top was the same in reverse. The actual run through of the cages occupied fifty eight seconds. Such organised activity for me was beyond the phenomenal.

I wondered, and I still wonder, just why my father had not tried his own fortune in a pit such as this one. Why had he chosen to scramble about such primitive organisations as Rough Lea and North Bitchburn? Just how reluctant to adventure could he have been?

It was the stabling of the ponies that astonished me.

For the first time in my life I found some endeavour to behave with humanity below ground. Those stables could not have been improved upon. The entrance was arched with the same careful craft as the tunnels through which I had passed. At the end of the arch which continued up the gentle gradient the stabling was organised along a passage running at right angles to it. A clean flow of air was borrowed from the main stream and passed through the accommodation for some forty ponies, all of which were munching the same old choppy mixture of the mines. The flooring throughout was cemented, and the stalls were separated by stoutly built brick walls. Light flooded all the stabling. The saddlery was hung across the passage behind the pony. All the walls and ceiling were lime-washed. The feeding boxes were copious and clean within the conditions of the place. There was no infestation of vermin. As a matter of fact rats were unknown in that colliery during the time I worked there. Mice existed in large numbers. The water trough was deep and clean, that is, so far as cleanliness can be observed in a mine. There were brooms and shovels to deal with the droppings. The runnels were kept brushed. Everything that could be done within reason was done for those imprisoned creatures. The horsekeepers kept a stern eye on youths and beasts. They examined each pony as it was returned to their keeping after the shift was ended.

I had not seen anything to compare with these underground stables. Obviously somebody did have some care for the little things. At Boulby the transport of the produced commodity was carried out by magnificent Shires and Clydesdales, which were returned after each shift to the stables near the edge of the cliffs like farm horses. But Boulby had been something special, and was a long way from Harton.

We geared our pony, a superannuated old fellow, who had been withdrawn from actual coal-putting, and was kept to do the odd jobs with the repairers on the night shift. We let him drink his fill at the trough before we set out to our destination, the landing in the Dandy Bank. The area supplied by this landing was 'coming back broken', that is, the pillars of coal which had been hewn many years ago were now being removed. This operation set free the pressures which had been contained over the years by the coal pillars. The space resulting from this extraction was called 'the goaf', and as this became enlarged by further extractions so the strata bore down against the obstructing coal and reached along the areas reserved for transport. Such illimitable pressure disrupted the ordinary life of the mine. The roof supports, even the great timber baulks set closely along the main transport avenues were crunched and snapped, and where this was held the floor of the mine was forced towards the roof. Consequently, the broken baulks had to be renewed and new ones inserted, and where the pressure could only 'find vent' by disturbing the floor and the tramway had to be taken up and relaid, and the debris carted away to some nearby pack.

Carting away was our task.

We arrived at the place of work and limbered the pony. This done we attached the limber lock to a small waggon, curiously enough called a 'kibble'. When this was filled with stone and rubble we made the pony drag it to the pack, the place, where it could be stowed out of the way. A great baulk, which had been summarily broken, was taken down and carried away, and when this was done the jutting stone was hacked down and a place cleared for the insertion of a fresh baulk.

The night passed slowly. I felt like a stranger who had strayed into a very strange land. The shift ended. We put the kibble out of the way near the deputy's kist, and led the pony back to the stables, where we unharnessed him, tied him up, and left him. Then we went to the shaft and joined the queue of men and boys waiting to be hoisted to the day.

As I left the cage an official drew me aside and told me that I would not be required to come to work until Tuesday morning at ten o'clock, and that when I did return I would ask for a Mr Simons and he would tell me what I had to do. I nodded acquiescence and walked on to the lamp cabin where I handed in my lamp, and out into the morning air. The air tasted sweet. The morning had yet to break. I walked away without haste. The air was blowing fresh from the seas.

Thus I began my new life in an underground from which I would not break free for thirteen long years.

Richard Llewellyn
First Day

Ivor came off the day shift and told me to get ready to go to work the next morning with him. I was in sweats with excitement to get my clothes ready, but nobody said a word in the house. Not a word. But the way they all said nothing, said more than if they had all climbed up on the roof to shout it over the Valley.

Next morning at quarter to seven I called for him, and my mother came with me as far as the door, but with no more fuss than if I had been going to school. I had my can, and my side pocket was heavy with five candles.

'Ready?' Ivor asked me, and Bron gave him his can.

'Yes,' I said.

'Well,' said my mother. 'Another one off, then.'

'Yes, Mama,' I said.
'Good-bye, now,' she said, and kissed me.
'Good-bye, Mama,' I said.
'Ivor,' my mother said, 'Look after him, now.'
'Yes, Mama,' he said. 'Good-bye. And good-bye Bron.'
'Good-bye,' said Bron, and a touch of a kiss for me.
And off we went, and my mother going quickly inside.

All the way down the Hill I had good mornings from the boys
and girls, all looking at me with smiles as though to say wait, you,
and you shall know you are alive in a couple of minutes. To the
men, of course, I was only another boy starting to work, so only
a few of them nodded, or gave me a tap on the back.

But going on to the pithead, I had the same feelings as when I
was in the boxing ring just before the fight was on. Something
moving in the belly, and heat in the head, and lightness.

Dai Bando and Cyfartha were coming running to get in the cage
when Ivor turned to go in, and me after him, looking back and
hoping they would reach us in time. The cage was a box made of
thick planks, bolted together on a steel frame, and the planks black
with years of use, and the floor inches in dust, and sounding like
a big drum.

Dai and Cyfartha squeezed in before the gateman locked up and
Dai saw me looking at him through the elbows of the man in front.

'O,' he said, and short with breath, 'you, is it? A bit of work now
then?'

'Yes,' I said.

And the ground fell from underfoot, and we dropped, with a
scream from the wind, into darkness, so dark that you thought
you saw lights, and your knees were loose and bent.

Hundreds of times I went down, but I never got over the drop
of the cage.

For moments you would swear you were blind. Then terror put
sharp teeth in you.

For hour after hour we seemed to be there, waiting, and the air
growing cold, but still dark, black, worse than night, and our feet
barely touching the falling floor, until it felt as though we were
standing in the middle of midnight with our knees bent ready to
jump into morning.

Then the scream dropped and dropped, and the floor came
firmer to the feet, the air was warmer and carried with it the salty
stench of raw coal, and light came to us, and breath and savour
of life to me, and gratitude, hotter than fire in me, for the gift of
sight.

'Come you,' said Ivor, when the gateman opened up.

I followed him through the arched brick of the pit bottom parting, and down the main heading that was noisy with trams and the singing of men working on them. The main heading was only wide enough for the trams to pass, with clearance for walking on both sides, and about nine feet high with lamps every few feet to give dirty yellow light.

We walked a good long way among crowds of other men until Ivor turned up a little hole in the wall, bent double.

'Come on,' he said, and smiling, 'mind your head.'

Up this pitch-black little tunnel we crawled, head almost to knees, and then Ivor stopped and threw his pick down.

'Right,' he said, and his voice coming like a roar in the dark. 'Light your candles, and I will show you what is next.'

So off with our coats and waistcoats and shirts, and I lit a couple of candles and stuck them, in their iron holders, into the prop. There was so little air that the flames went to six inches with them, and pretty indeed.

'Now then,' Ivor said, 'I will cut the coal, and you will push the lumps down the chute. Then go down and load all you find down there into my tram, is it?'

'Yes, Ivor,' I said.

'Right,' he said, and his pick punched deep into the seam.

So I started to work.

Ivor was a good workman, quick with his pick, untiring, and stopping only to move slag that fell when the coal was loosed. When he stopped, I stopped, but not to stop altogether, for we banked the slag against the sides and packed it tight to act as a prop for the roof.

For hour after sweating hour, bent double, standing straight only when we were flat on our backs, we worked down there, with the dust of coal settling on us with a light touch that you could feel, as though the coal was putting fingers on you to warn you that he was only feeling you, now, but he would have you down there, underneath him, one day soon when you were looking the other way. I used to look at the shining black strip in the orange light of our two candles, and think to myself that this might be the mourning band of the earth, and us taking it from her to burn, and she looking at us with half-shut eyes, waiting to have a reckoning. But there was always a fear in me, down there, that I never lost.

I always seemed to hear a voice in the heavy quiet, beyond the punch, punch, punch of Ivor's pick, and the rolling echoes of coal sliding down the chute. And I always thought I saw a face in the glitter of the coal face, and never mind how much Ivor cut from it, it always seemed to be there.

The muscles of the belly might feel to be tearing apart long before the end of the day, so bent we were. Ivor would kneel, lie on his side, stand sideways and bent, or on his back, with sweat making his skin into black silk, but never a pause, never a stop, till it was time for eating, or for a swill of tea to take dust from the throat.

I knew well, even on the first day, where Dai Bando had those muscles in the belly.

And, O, what joy to come up in the cool air of night after hot hours in the light of candles, light that crawled with dust that sometimes shone. Then I knew, and knew with thanksgiving, why we sat on doorsteps when the sun was out. Only to be quiet, and rest aches, looking at clean light, feeling the blessing of the sun, free, for a couple of hours, from the creeping touch of the fingers of coal.

Up the hill, among the crowds on the shift, and passing boys I knew without a nod from them, and surprised, until I remembered the top skin of coal dust that covered me from head to foot and hid me from them.

But I felt a man in real truth, to be coming up among that crowd of men, sharing their tiredness, blacked by the same dust, knowing the sounds and the sights of the colliery as they did, thinking with the same mind, of them, with them, a part of them.

I bathed with Ivor in Bron's back, for there were more than enough in ours already.

There is good to see the tubs ready and the buckets all lined up, steaming. Off with the clothes and leave them where they fall. One bucket over you to take off the worst, then a rub of soap, another bucket, more soap. Now you will see a bit of yourself, but the hands, and especially those little lines in the balls of the fingers, are hopeless. You shall scrub and scrub, but Mr Coal will lie there and laugh at you. A good friend to man is water, indeed, but never friendlier than when he is running down your back, chasing coal dust off with a stick of soap.

Into the tub, then, to rub a white lather all over you and duck under the water, holding breath to feel the gentleness all round you, close as your own skin.

Idris Davies
When You Were Young, Dai

When you were young, Dai, when you were young!
The Saturday mornings of childhood
With childish dreams and adventures
Among the black tips by the river,
And the rough grass and the nettles
Behind the colliery yard, the stone-throwing
Battles between the ragged boys,
The fascination of the railway cutting
On dusty summer afternoons,
And the winter night and its street-lamps
And the first pranks of love,
And the deep warm sleep
In grandmother's chapel pew
On stodgy Sunday evenings,
And the buttercup-field you sometimes noticed
Behind the farthest street, the magical field
That only the heart could see,
The heart and rarely the boyish eye,
And the pride you had in your father's
Loins and shoulders when he bent
Between the tub and the fire,
And the days you counted, counted, counted,
Before you should work in the mine.
You never, never cursed your luck
Or desired to see another town or valley,
Or know any other men and women
Than those of the streets around
The street where you were born.
Your world was narrow and magical
And dear and dirty and brave
When you were young, Dai, when you were young!

Lewis Jones
Len Starts to Work

One day Big Jim came home from his repairing work at the
colliery, and after dinner casually told Shane: 'Well, gel bach, we
will be starting to fill coal in the old pit next week once again.'

Shane turned to him. 'That be good news,' she said.

'Ay, but I have got even better than that. I have arranged with Williams, the under-manager, for our Len to start work with me next week.'

Len sat up in his chair with a start and looked at his father with wide-opened eyes that sparkled with interest.

'Aye, aye. It be quite true. You be starting to work with me next week. Your mam wanted you to keep in school, but since you be not willing for that there be nothing left but for you to work.'

The week following this brief announcement Len was so excited he hardly knew what he was doing. He told his mother one day: 'I'm glad, mam, that I'm starting to work. School's no good to me, I can't learn enough there. I want to be with dad in the pit. I'm not afraid. The other boys have told me it's not so bad when you get used to it.' And, looking slyly at his mother, he added: 'They get pocket-money from their mothers on top of trumps from their butties.'

Shane pretended not to notice, but a tiny smile flickered for a moment on the corners of her mouth.

Len became a hero in the eyes of his schoolmates. He made them envious with tales of what he intended to do, the things he would buy, the places he would visit when he started work. He conjured up for them a romantic vista of what work meant, and the days went by so slowly he thought the week would never end. But at last the final night arrived.

His mother had bought him the usual white-duck trousers that marked the end of his boyish breeches. The large tin box and jack, to carry his food and water, were given to him by his father, who had already used them for years. Shane sent him to bed early, intending him to get plenty of sleep. Excitement and anticipation, however, prevented this and he was still awake when the first morning hooter blew at five o'clock. About ten minutes later his mother called him. She had already lit the fire and had breakfast waiting. Big Jim was half dressed in his pit clothes when Len entered the kitchen. The lad soon clothed himself in his strange rig-out, and sat down to drink a cup of tea, for food was out of the question in the state to which he had worked himself. When the half-past five hooter gave the signal that it was time they were off, Shane passionately pressed her son to her body. She kissed him tenderly and whispered in his ear, 'Do everything your dad do tell you, my boy. Don't move from his side. You be starting to-day what only the grave can steal you from.' She said this more to herself than anyone else and put the canvas apron to her eyes.

Then shaking her head sharply, she turned to Jim.

'I know you will take care of him, James. 'Member he be only a baby after all, the only one us have got left.'

With another hungry kiss she sent them into the dark, wet street with a 'Good morning' that stuck in her throat and was never uttered.

The rain poured down as Len and his father, like a giant and a pigmy, trudged up the hill towards the pits. Jim made his son walk as closely behind him as possible. It was some time before Len realised this was done to shield him from the main force of the rain driving down the valley. His head bent to the drops that evaded his father's body, Len vaguely noticed the long string of silent men, like shadows, making their way in the same direction as himself. Each of them, dragging his feet, used the man immediately in front to shelter him from the rain.

Len followed his father across the bridge into the cabin where authoritative-looking men scanned him over curiously as if he were a calf.

'So this be the boy, James?' asked the most officious.

'Ay, this be him,' responded Jim, whereupon Len had to write his name and age in a big book.

He took a round piece of metal handed him by one of the men, who told him, 'Take care of that, my lad; it is your lamp check with your number on it.'

A few further words passed between the group before Jim led the way to the lamp room, a long corrugated-iron structure containing hundreds of lighted lamps arranged on a series of trestles. Len followed his father to one of the pigeon-holes, where Jim handed in the check and received a lamp in exchange. Len did the same and at once felt himself a man, although the lamp dangling from his hand nearly touched the floor. They left the lamp room and walked to a cabin, where another man with an air of authority examined the lamps. Having unscrewed the top and blown all round the pots, he handed them back with a final twist of the bottom, to ensure that the lamps were stuck fast.

Big Jim and his son left the cabin and went straight to the pit-head. The shaft in which Len was to work was called the 'upcast,' because all the air from the pit was sucked up through it by a fan of huge dimensions. To prevent the air being drawn up before it had time to circulate all round the workings, the shaft was closed in with heavy wooden 'droppers,' only leaving a space for the rope to wind its way through, so that when the two cages were in the pit the air howled and screamed through this tiny outlet.

Len was rather frightened by the terrible tumult on the pit-head, and he had to shout to make himself heard above the din. While they were standing in the queue waiting to go down, he felt for his father's hand and pressed it to his side in a gesture of love and confidence, but Big Jim, sensing the boy's mood, said nothing, thinking to himself that the lad had to 'find his own feet'.

When the ascending cage lifted the wooden barriers from the pit-top the released gusts of heated air rushed through with a roar. Jim cautiously led his son over the little gap between the cage and the pit edge. Eighteen other men and boys followed before the man in charge declared that the box was full and placed a thin iron bar across the entrance. This was a measure of precaution supposed to prevent the men falling out.

Once inside the cage, Len held his breath and waited. He heard the knocker clang three times, and the tinkling of a bell far away in the engine-house. Then suddenly he felt the floor of the cage press against his feet as it lifted off the stanchions that held it to the pit-head, and in another second the breath was torn from his lungs by the sudden drop as the cage plunged its way into the depths of the pit.

Even in his panic Len heard the clatter of the droppers falling into place above him, and he felt that a door had been bolted between himself and the world. Regaining his wind after the initial shock he put his arms round his father's leg, finding courage in the human contact it provided in the black, falling void. Warm air rushed past the cage with wicked squeals, and just as Len was beginning to get accustomed to the sensation of dropping, the bottom of the cage again pressed against his feet. This was due to the brakes in the engine-room being applied to the great winding drums and marked the half-way line between the pit bottom and the surface. The lad felt the cage rising under him and wondered why they were returning to the pit-head, but before he had time to think it out the cage, with a few preliminary jerks, jarred on the planks that covered the water sump at the bottom of the pit.

The men slowly got out, Len behind his father. His curious eyes noticed at once that the little lamps appeared to give out a greater light here than they did on the surface, due to the more limited space they had to illuminate. He was also surprised that he could see underground as well as he could above, for he forgot there was no daylight when he left the pit-head and that his eyes were already inured to the darkness before he descended. He stumbled against a rail and glanced around the semi-elliptical passage-way that led from the pit bottom. Looking behind, he saw a similar passage going in the opposite direction, the other side of the pit, then, threading his way carefully between the long line of coal-laden trams on the one side and an equally long line of empty ones on the other, he eventually came to the end of this double roadway.

'Look where you be going to now, Len bach,' advised Big Jim, as they turned off into a narrow and gloomier passage.

'There's different this place do look, dad, without no whitewash on the sides.'

'Never you mind about the sides. You watch these ropes in the roadway.'

Len took his father's advice and kept his eyes glued upon the ropes. Not another word was said until they came to a large cabin dug into the side of the roadway, where their lamps were again taken from them and tested.

Big Jim received certain instructions from the man in the cabin, and Len listened to their talk of 'shots' and 'rippings'. After receiving instructions and having their lamps returned to them, Big Jim remarked to Len: 'Come on. We have got a hell of a plateful for to-day, so you'll have to look sharp.' Without a word Len followed his father. The roadway was becoming narrower and lower with every stride, and steel girders gave way to timber as supports for the roof.

In his anxiety to keep pace with Big Jim Len had no time for talking; it was his father who broke the silence.

'Are you all right, Len?'

Len started at the sound of Jim's voice. 'Ay. I'm all right, dad. How much further have we got to go?' he asked tremulously.

'We got a good bit to go yet.'

'Have we come two miles so far?'

'Thereabouts, boy bach,' replied Jim, turning to Len. 'You don't feel tired, do you?'

'Of course not,' bragged the lad, forgetting the tired ache in his legs and pulling himself to his full height.

Jim walked on again. 'Many is the time I have tramped this old roadway,' he mused. 'Duw, duw, I be sure to have done hundreds of miles along it.'

They proceeded in silence for a while, Big Jim thinking of the good old days that had gone, while Len thought of the days that were to come. Suddenly the former warned: 'Look out by here, Len bach. This be a nasty old trip.'

Len continued down the steep road that reminded him of the path leading from the mountain top.

'I 'member coming over this trip once with old Dai Cannon,'

said Jim, 'when all of a sudden we did hear a rush behind us. Dai
and me stopped like statues for a minute. But not for long. The
rush come louder and louder and, muniferni, you did ought to see
us jump for the side. Ha-ha! Dai gived one howl, mun, and before
I did know what was happening he fell back on 'is arse in the
middle of the road. Ha-ha-ha! Arglwydd mawr, boy, I was bound
to laugh, mun, if Dai did kill me for it. Arglwydd, you should have
seed the look he did give me! "That's right," he did say, "laugh
you silly beggar. Go on. Enjoy yourself although I have broked
my bloody neck."'

Jim burst out into another uproarious guffaw at the memory.
When this had subsided Len asked: 'But what was the matter,
dad?'

'Why, some of the horses, coming down from the top of the trip,
got a little bit restless and was stamping their feet like hell. 'Oops
a daisy, we did think, the devils be running wild. Dai jumped for
the side and bumped right into a low piece of timber, and that was
why he did land flat on his arse. Ha-ha!'

The descent now became even steeper. Len compared it to the
sheer mountain drop near the quarry, and the thought made him
long for the first time that morning to be back again in the sun.
He had never dreamed of this interminable tramp in the darkness
of the pit. Thinking of the world above prompted him to ask: 'How
far be we down, dad?'

'They do say 'bout two thousand feet. But never mind 'bout that,
now. You look after yourself; the roof be getting pretty low by here.'

Jim was walking with his body bent nearly double and Len,
taking the tip, dug his chin deep into his chest and bent his head
low. After a while they came to a part of the road where the roof
was higher. Jim, knowing the spot, straightened his body and
walked erect, but Len, fearing to raise his head, was not aware of
this and walked on in solemn silence with his head bowed like a
man in a funeral. The ropes beneath the lad's feet were moving
when Jim called out, 'Come into this manhole, boy bach.'

Len hurried into the tiny hole in the side of the roadway and
squeezed himself alongside his father in the limited space. 'What's
the matter, dad?' he asked, thinking that something serious was
about to happen.

'It's only the journey,' Jim assured him. 'You see those ropes by
there,' pointing into the roadway, where the ropes were slithering
along like snakes, one sizzling along the ground while the other
ripped through the air nearly to the height of the roof, 'Well, those
do belong to the journey'.

Len hesitated a moment, then asked, as a low rumbling sound came to his ears from the distance, 'What's that noise, then, dad?'

Jim started to explain when thirty empty trams rushed past the manhole with a deafening clatter. The terrific din sent Len cowering against his father's legs. Without further explanation, Jim caught the lad by the arm and drew him out of the hole. 'Come quick,' he shouted above the rattle of the receding trams, 'let us get to the parting before the full journey come out.'

Len did not understand the meaning of these instructions, but he obediently ran headlong after his father. Gasping and perspiring, they stopped after they had run about a quarter of a mile. Still panting from his exertions, Len noticed that the roadway had widened and was blocked by two strings of trams, one of which was empty, the other full of coal.

Safe in another manhole, the lad watched with interest two men change the rope from the string of empty trams and place it on the full ones. When the change was completed he heard a whistle blow further on, and one of the men near him responded to the signal by rasping the two thin wires above his head with the blade of a knife. The wires connected with the engine-house at the bottom of the pit. The ropes began to move and slowly tightened on the first tram of the string, then the others, attached to it with steel shackles, were drawn forward with increasing speed and clamour until the last was lost to sight in the darkness of the roadway.

Len and Jim emerged from the manhole and again continued their walk, the former beginning to think that it was to be endless. He noticed places where huge holes gaped in the roof. At other places he saw large masses of stone overhanging into the roadway without any visible support.

He turned to his father and asked in a quavering voice, 'Be that safe, dad?'

'Safe? Ay, boy, safe as houses. It will take more than Gabriel's trumpet to blow that down.'

Len was too fatigued to ask any further questions. He was glad when Big Jim stopped and said, 'Here we be. Strip off and get yourself ready. A little whiff will dry up all that sweat on you.'

Len's exhaustion vanished with the knowledge that the interminable trudge was over and that he was now in his father's working place. He started to pull off his coat, when Jim interrupted him testily. 'Not by there, boy bach. Shift under those timbers, where you will be safe.'

Len did as he was told, and putting his box and jack carefully at the foot of a strong-looking prop, he pulled off his coat and shirt. He paused at this until he saw Jim pull off the singlet next his skin; then he did the same, and immediately felt the air beat more cool and pleasantly upon his naked chest.

'Duw, that be nice, dad,' he said, revived by the contact.

'Huh,' grunted Jim. 'Take the tools off the bar. Here be the key.'

Len did so, then, with a shovel in his hands, he followed his father on hands and knees through the coal-face. The glistening coal, reflecting the gleam from the two lamps, fascinated Len. He watched Jim crawl, practically on his stomach, up the long stretch of the coal face until only the dim light of his lamp was visible. Scared to be left alone, the lad followed, only to be gruffly ordered back.

'You keep by that empty tram and don't move till I tell you.'

Len turned back and for the first time gave conscious thought to the tram. It stood end on to the clear-cut roof, or 'rippings,' which had to be blown down as the coal-face advanced, so that the tram could follow the coal.

A deep feeling of loneliness enveloped Len as he wondered what would happen if his father were not near and he were left entirely on his own. All round him he could hear little movements, as if the place were alive. He had an uncanny feeling that the roof was moving, and each creak of the timbers, as they unwillingly took the weight of the settling strata, sent a quiver through his body. He had yet to learn that the pit had a life of its own, that it was never still or silent, but was always moving and moaning in response to the atmosphere and pressure.

Suddenly he felt a burning sensation on his stomach. His hand flashed to the spot automatically, his fingers clutched some object and tore it away, and opening his hand he saw a huge red insect with innumerable hairy legs and hard, shiny wings. Although crushed in his convulsive grip, the ghastly legs still beat the air, and looking down at his belly, he saw a thin stream of blood running down it where the cockroach had gripped the flesh and torn it away. A sick giddiness swept over the lad for a moment while the perspiration burst from every pore in his body lathering it in a mixture of coal-dust and moisture, but before he could recover from the shock he heard his father crawling back. This proof that he was not alone encouraged the lad and he was smiling when Big Jim emerged on the roadway.

'We will work in the right hand cut to-day, Len bach,' he said, 'so that we can free the whole face for to-morrow.'

Len did not understand the technique underlying the remark, but he asked with assumed indifference, 'What be I to do, dad?'

Jim replied: 'You will come up the cut with me and throw the coal back towards the tram.'

The lad obeyed, and followed his father, and for hours he worked on his knees with the back of his head rubbing against the roof. He began mentally counting each shovelful of coal his father cut and which he had to throw back to the tram. His arms grew heavy as lead, cramp caught him in his bent legs, and his back felt as though it were broken. The coal-dust that filled the air got into his nose and eyes. It made him sneeze and blink and, working into the sweat-opened pores of his body, set up an intolerable irritation. He felt it impossible to lift another shovelful of the coal he now detested, but somehow he kept on, until at last his father said:

'That will do for now. Let's go back and get a bit of tommy.'

The lad dragged his weary, painful limbs back into the roadway, where he stretched himself full length in the dust. He saw his heart pumping against the bones of his naked chest, and felt pins and needles run through his flesh in spasms of excruciating agony.

Big Jim, sensing what was happening, urged him to his feet. 'Come on. Get up before you go stiff.'

With infinite care Len dragged his limbs together and slowly rose to his feet. He opened his food box and sat down. The bread-and-butter looked dirty and unappetising, but the water in his jack was like nectar. Jim stopped him before he had emptied the tin of its contents. 'Don't do that again or you will get cramp in your belly. Get on with your food.'

The lad tried to obey, but the hundreds of savage-looking cockroaches that buzzed and fussed around turned his stomach, while the dust he had already swallowed curdled in his inside.

After a while Big Jim rose and made his way back up the face, telling Len: 'You stop there till I shout for you. A bit of a whiff 'on't do you any harm now.'

During the rest the lad slowly recovered from his exhaustion. The black dust under his body seemed softer and more sweet to him then than even the green grass on his beloved mountain, and his mind wandered to the end of the shift. Before his eyes floated a picture of the envious glances of his schoolmates when they saw him striding, black-faced, down the hill in his working clothes. He saw the glad look in his mother's eyes as he walked into the little kitchen, having finished his day's work. Already he began to count the pocket-money he would have in a fortnight's time, and speculated how best to spend it.

Deeply immersed in these pleasant contemplations, Len dozed off into a heavy sleep. Jim's deep voice seemed miles away when he shouted, 'Right you are, Len bach; come up and start chucking this coal back.'

Len came back to reality with a start and made his way up the coal-face he already hated with every fibre in his body. He worked in a semi-conscious state, only faintly aware of the three or four occasions when the haulier and his horse noisily changed the full tram of coal for an empty one. When the fireman came round and chatted with Jim he waited respectfully on his knees, wishing fervently that the man would stop there till the end of the shift. But he had ceased to take any interest in what was happening. His brain was numbed with the physical exhaustion that again consumed him even though his father had been careful to limit the amount of work to a minimum.

The poor lad, accustomed to the fresh air of the mountain, felt the foul atmosphere of the pit beginning to choke him. He thought again of his mother, and now wished he had listened to her advice and tried the examination for the secondary school. Young Mary, Ezra's daughter, had done so and passed successfully, although she was no better scholar than he. Too late now, he thought to himself, half weeping; now he had started in the pit he had to continue. He wondered if the rain had stopped up above; it seemed years since he had left its refreshing coolness. He was sorry now he had ever grumbled at the rain, and was willing for it to pour down for ever as long as he was on the surface to see it.

Tears involuntarily gushed to his eyes and he was on the point of bursting into sobs when a terrific crash shook the whole earth. For a moment he stood paralysed with fear, then he rushed headlong with a wild scream towards his father. Big Jim caught the terror-stricken, hysterical lad to him.

'Duw, duw, mun, don't ever let it be said that the son of Big Jim is frightened by a noise. That was only gas and squeeze busting inside the coal, mun. There, there, now, don't be 'fraid no more.'

It took Len some time to control his quivering flesh. The crash had sent the memory of the explosion flashing through his mind, and in a split second he had seen himself in the place of the bodies he had watched being buried on the day of the funeral.

Jim made the lad rest back on the roadway again until, some half-hour later, he took him down the roadway to fetch some timber. Here Len saw another lad with his adult mate engaged in the same task. The sight of someone his own age immediately restored his confidence. His natural taciturn unsociability evaporated with the new contact in the new environment. While the two men were chatting and selecting the timber they wanted, Len shyly asked the strange lad, 'How long you been working?'

'Oh,' was the casual, off-handed reply, 'a long time now, butty. More than six months, I believe, though I can't 'member 'xactly, because it be so long ago.' Saying which, he took a lump of chewing-gum from his mouth and spat noisily into an empty tram near him.

'Well, what do you think of the bloody hole?' patronisingly.

'Not so bad,' lied Len, trying to forget the torments of the day.

'Huh. I'm only sticking it till I'm old enough to get a horse.'

'Get a horse?' queried Len amazedly.

'Ay, ay, that's it. I'm going to be a haulier.'

'Oh, I see. Like that man who do bring the horse to fetch our full trams out?'

'You got it, butty. And, by Christ, can't I handle them!' warming to his subject. 'Take a tip from a old hand, butty, never take no bloody nonsense from them. When they turn twp or stupid, a sprag will always bring them to their senses.'

He accompanied these remarks with a clicking sound and a practical demonstration which left Len staring with admiration. The budding haulier put the chewing-gum back into his mouth with a grimace and remarked, 'This bloody stuff be getting too weak for me now; I will have to start chewing 'bacca soon.'

Len felt he would like to have a say: 'I only started to work to-day,' he said hesitantly.

'Be that so? Ah well, never mind, you'll soon get used to it when you have worked so long as me.'

Their conversation was interrupted by the man with Big Jim.

'What the hell be you blabbing 'bout by there?' he demanded. 'Why don't you come and give me a hand with this blasted timber?'

'All right, all right, keep your wool on,' the lad said casually, and turning to Len, he hurriedly whispered: 'That's my butty. I 'spose I'd better go and give him a hand. Come out same time as me to-night — we be working next place to you. My name is Will Evans. So long.' With this he and his mate left, each with one end of a long nine-foot prop on his shoulder.

Shortly after this finishing time came and Len gathered all the tools together, his father showing him how to put them securely on the tool bar.

He dressed in quick time, the clothes sticking to his steaming body, and as he envisaged his triumphant entry into the house all his old pride began to surge through him again. To make sure his face was quite black he rubbed it vigorously with his dusty cap.

On the way out he told his father of the request made by Will, the lad in the next working place. Big Jim took him round to it, and they both waited for the others to finish; then the two men and their boys went out together, the former in front.

Len felt elated as he retraced his steps along the roadway that in the morning had seemed like the pathway to hell. He chattered incessantly and already felt he was an old hand in the pit. His new-found mate let him ramble on for a while, then broke in with the question: 'Do you know Sam Dangler?'

Len shook his head negatively.

'Huh, you have missed a treat. That's a haulier for you, mun. You ought to see him handling the rough 'uns.' He stopped half-way up the trip. 'This is how he do do it,' he remarked, catching hold of an imaginary rein. 'Whoa boy, whoa! Ah, bite you sod, would you?' giving a sharp tug and pressing his body back as though he were pulling the non-existent rein. 'Take that, you bloody cow!' hitting the air with a short piece of timber. 'Whoa, boy. Steady now. Ah, that's got you. Come to your senses, have you?' He flung the sprag into the roadway triumphantly and remarked: 'That's how Sam Dangler do conquer um, see?' A moment's pause, then: 'He's a devil. All the horses do know him, and after a week he have very near got them talking. I'm going to be like him one day. He do have more trumps off the colliers than any haulier in the pit.'

He broke off here to take Len on the side and whisper in his ear: 'You want to watch your old man on pay day. Tell him straight from the beginning that if he want you to work he have got to give you trumps. Huh. Fathers be the worst butties going. They do think their own sons be bloody slaves and do never think of trumping 'em. Oh, no. They do pocket that their bloody selves and the old 'ooman don't have a smell of it. You listen to me,' he continued sagely. 'Don't let any butty make out of you unless he pay you for it, father or no father.'

A voice drifted down to the two lads from the top of the trip: 'Will-o! What the bloody hell be you hanging like a shirt behind there for?'

They hurriedly continued their way, and eventually overtook their mates on the pit bottom, where they had to wait a while in the queue before they were bundled into the cage.

The sound of the iron knocker, announcing to the men on the surface that all was ready, came to Len's ears like the chime of sweet-tolling bells. The cage sprang up the shaft like a projectile released from a mighty catapult, and in a matter of breathless minutes the roar of the air beating against the droppers drowned every other sound. There was a clamouring rush as the chains caught the covering and Len once more saw the light of day.

Although it was not a bright day, the light hurt Len's eyes, and he had to close them for a while. Will waited for him, then led the way to the lamp room, where they exchanged their lamps for the numbered metal discs. Having done this, Will led the way to some iron girders near by and began poking his fingers into the crevices. His movements became more vigorous and excited and his face grew red.

Len noticed this and asked timidly: 'Be there anything the matter, Will?'

'Matter? Matter to hell!' exploded Will. 'Some dirty swine have pinched my fag. I put it by here this morning before going down, and now the bloody thing is gone. Ah,' he grunted viciously, twisting his cap in his hands, 'if I only had him by here now I'd wring his dirty neck.' He gulped something down his throat and turned to Len. There was a pathetic look in his eyes when he said, 'Butty, that is the worst thing a man can ever do. To pinch a fag is worse than pinching grub. Huh. Hanging be too good for a man who do a thing like that.'

Some other lads now approached and listened interestedly to Will's tirade against the unknown thief. One of them eventually handed over a piece of cigarette that he had been smoking himself, and the little party made their way homewards. Seeing all the other lads enjoying their cigarettes made Len feel out of place, and he decided he must learn to smoke as soon as possible.

He took his place in the long line of men that streamed homewards down the hill, and looked surreptitiously into every window to see the reflection of his coal-blackened face. The eyes which looked back at him seemed twice their usual size in the dark frames.

When he reached home he found Shane busily fussing around with the dinner. Len never felt so proud in his life. Pulling off his coat, he put it in the box under the table and jovially remarked, 'Hallo, mam. Here I be, safe and sound.'

Shane did not look up from her task as she answered in a voice that quivered a little: 'Ay, I see you are, Len bach, thank God. How did you like it?'

Len forgot all the pains and terrors of the day when he replied, 'All right, mam. It be not so bad and I have made a new butty.'

He proceeded to retail the events of the day. When he had finished Shane beckoned him to wash his hands in the bowl of water she had ready. This done, he drew his chair to the table and voraciously ate the kipper and potatoes she had cooked him.

He had nearly finished his dinner when Big Jim came in, and explained to Shane he had called in to have a pint of beer to clear his throat. He always claimed, and believed, that if he did not have this pint of beer he would be unable to eat his dinner, because of the coal-dust that coated his lungs.

After dinner Shane fetched in the wooden tub made out of half a big barrel. Jim lifted the boiler from the fire and poured its contents into the tub, while Shane cooled it with cold water. Len quickly stripped to his waist while his father enjoyed a smoke, and was thrilled at the fact this was the first time in his life he had bathed with the top half of his body over the tub.

While he wiped his head and shoulders Jim followed him into the tub; then, when the latter had finished, Len got in naked to wash his bottom half. He enjoyed the sensation of the warm water running down his legs and lingered in the tub until Jim said: 'You be in there long enough to swim round it, boy bach. Come out and give your father a chance.'

Big Jim had hardly started rubbing soap into his legs when the door opened and Mrs Thomas, their neighbour came in.

'Can you lend me a bucket of coal till mine do come from the pit?' she asked, taking no notice of Jim's magnificent nude body.

'Certainly, gel fach, go out the back and fetch it,' he replied, completely unabashed by her appearance.

By the time Mrs Thomas had filled her bucket of coal Len had dressed, and when she re-entered the kitchen he was peering at his face in the cracked mirror. The two black rings of coal-dust that circled his eyes made them show up vividly against his white face, but though he rubbed them with the towel until the rims were sore, he could not remove them entirely, later in the evening when he walked through Main Street he was proud of the black rings, because they showed to all and sundry that he had started to work.

Anon.
The Plodder Seam

The Plodder Seam is a wicked seam,
It's worse than the Trencherbone.
It's hot and there's three foot of shale between
The coal and the rocky stone.

You can smell the smoke from the fires of hell
Deep under Ashton town.
Oh, the Plodder Seam is a wicked seam,
It's a mile and a quarter down.

Thirteen hundred tons a day
Are taken from that mine.
There's a ton of dirt for a ton of coal,
And a gallon of sweat and grime.
We crawl behind the cutters and
We scrabble for the coal.
Oh, I'd rather sweep the street than have
To burrow like a mole.

Bob Smith
It Was a Man's Life

My mother had to waken me the next morning, shaking me gently
and speaking quietly so as not to disturb my brother Wullie. I had
been sleeping so soundly that I had not known he had joined me
in the bed. 'Is it that time already?' I asked, drowsily. 'I could sleep
for a week.' And surely I felt as though I could, but it was another
day, and I was a collier, and had to be up and away. My breakfast
was waiting, and Father was sitting by the fire lighting his pipe.
'Coom on, Son, dinna be late or we'll miss the bus,' was his
greeting. 'Ah could sleep for a week, an ahm sore,' I told him.
'Aye,' he replied, 'Ye will be, but the soreness will go in a day or
twa. Hurry up an git yer claes on.'

Mother had put my pit clothes down by the fire to warm, for
it was a day of hard frost. Already the clothes had that musty
smell of the pit about them, and my boots, brushed, cleaned
and dubbined by Mother, felt as hard as blocks of wood.
'Tackety bits', we called them, and they were heavy and strong,
for they had to protect our feet, but they took some getting used
to.

Father was ready to go, and called me to hurry. I suppose I was
still sleepy, and certainly my dreams of that first shift were still
with me. In fact, this was a decisive moment for me. I could have
said that the mine was not for me, and gone back to my bed. That
was what some of the older men had hinted yesterday, and that
was what my Mother, at least, hoped would happen. Was I really

going to be a miner? Was I really going to spend all my working days down a pit? Was I going to join those hunched, crouching figures as they hurried down the underground roads to their working places? Was I going to spend years in that awful atmosphere of stench and dampness and the acrid reek of explosives? Was there to be always that hellish noise of the haulage ropes hauling their rakes of hutches along the main road? Were my days to be divided into sections by the rattle of the pony chains as a new hutch was brought along, another one to be filled, another one in an endless succession of them? I stamped my feet more firmly into my boots, picked up my piece tin and flask, my lamp and carbide tin, and went out of the door after my father. I was going to be a miner, like him.

It was still dark when we reached the pit head, and the few lights showed the way up the rickety steps to the gantry. It was a stark scene in black and white, with a few columns of steam rising into the frosty air from leaky joints and valves on various pipes. There was no colour, nothing natural, nothing growing there. This was industry, the business of wrenching coal from its ancient bed, and it was very far removed from the village and the woods I knew so well. This was where my father and all the men I knew spent a lot of their lives, and I was now one of them.

My father sent me off to the checkweighman's hut to pick up our pins. The checkweighman was a very important person. He was elected and paid by the men themselves, and it was his job to check the weight of every hutch that came up the shaft. Of course, the management also weighed the hutches, but in the past it had been clear that the men were not always paid for the coal they had won. So the system of appointing checkweighmen was set up, and he ensured that the men were credited with all the coal they sent up. Each man had a set of pins, metal discs stamped with their individual number, and a pin was put on every hutch before it left the work place. At pit head each pin was removed, and the number noted, and so every man was properly credited with the work he had done. At least, he was now that the checkweighman was present. I went up to the bolthole in the checkweigh hut and called out '88' — that was Pop's number, and it appeared on all his pins. McLean, the checkweighman, put his head out of the bolthole. 'Hello, Son, an hoo dae yer like the pits?' I told him that I thought it not so bad, and he replied that it was a hard life I was letting myself in for. 'I ken yer faither, an he's a guid man. If yer as guid as him ye'll dae,' and he passed out the pins. I went back to the men waiting their turn to go underground. 'Wha's last?' I called for the first time, again with a feeling of pride that I was a man amongst men.

There was thick black smoke with a strange smell hanging around the pit head, and I enquired what it was. 'What's aa the reek?' I said. 'That? That's the pit heedman pittin his sweety pitsocks on the brazier,' replied Big Tam jokingly, and explained that braziers had been kept burning all night round the pit head to keep the shank free of ice. 'It's been a gae hard frost through the nicht, an the cage'll be like an ice-box this mornin. Mind an watch yer feet,' said Big Tam, and explained that the cages had been kept running up and down the shaft all night to keep it free from ice. If that had not been done, and there had been icing up of the shaft, we could not have gone down that frosty morning.

I was near the front of the line by now, and waiting my turn. The winding engineman's signals rang out loudly, and a cage dropped from sight down the gaping hole of the shaft. The wire rope hummed, and then hissed through the cold air as the winding engineman opened his valve. Shortly there was a low rumble and then a crash as the other cage came into view and stopped. The onsetter leaped forward, opened the gates, and grabbed the rings of the full hutch. He kicked off the guard that held it in place, and pulled the hutch forward, then jumped to one side as it went off to the tumbler. He rang 'Man Riding', and we entered the cage. It was full of reek from the braziers that were burning at the pit bottom, and today seemed even more like the entry into Hell. There was thick ice, two or three inches of it, coating most of the cage, and we had to tread carefully over the angle irons that served as rails for the hutches. The hand rail was also covered in ice, and wee Charlie Fisher took off his scarf and wrapped it round the rail before taking hold. 'Canna get cauld hands. The wife disnae like them at aal.' There was laughter at his joke, but most of the men followed his example. I simply pulled down the sleeves of my over-size jacket and grasped the rail like that.

The onsetter gave the signal to go, and it felt as though the bottom had dropped out of the cage. I gasped for breath and felt real fear. Charlie noticed and laughed. 'Dinna worry, lad, there's a rope haudin on tae the cage, Ah hope. If it broke we'd git tae oor wark all the quicker.'

'Shut yer face, ye daft bastard,' shouted Big Tam. 'Don't scare the wits oot o the laddie. He's no used tae this, so shut that daft gob o your's or ah'll shut it fer yer.'

There was no more talking amongst us. The cage was shuddering and jerking, and everyone, including the experienced men, was uneasy. Lumps of ice rattled down onto the roof of the cage and we huddled closer, covering our faces with a free hand. 'Gees,' someone said, 'Ah wish we were oot o here.' 'It'll no be long noo, Son,' Big Tam said to me, as the lights of the bottom appeared. 'Anybody hurt?' he enquired, as the cage gave a final lurch and stopped at the bottom. 'Thank Christ we're doon,' he added quietly to me. 'Were ye feart?' 'Na, na, Tam,' I replied, although my white face told another story. 'Weel, ye'll get mony a feart in the pit,' he said.

There was a shout of 'Haud on, we hive an injured man.' Charlie was leading Rab, who was holding his hand over his face, with blood seeping through his fingers. 'Auch, it's no sae bad. Ye kne ah hardly felt it,' said Rab, unwilling to make a fuss. 'Ah thocht it was a splash o watter as we passed thon wet bit.' Sandy, the pit bottomer, lifted Rab's hand away from his face and examined the wound.

'Christ,' he said 'Ye've given yersel a right nast yin. There's a big cut doon the side o yer face. Haud on and ah'll git a rag and stoap that bleedin. Ah'll fix up Rab,' he added to the rest of the men, 'an ye can get oan to yer wark.'

We left Rab to the rough ministrations of the bottomer and made our way to the fireman's cabin. Father was waiting for me there, and I told him about Rab's accident, that a lump of ice had fallen down the shaft and cut his face open. 'Damn it,' said Pop, 'there's always something,' and went on to shout 'Right, Jock,' to the fireman. Jock marked his book, recording Father's presence down the pit, without even lifting his head, but when I also called 'Right, Jock,' he looked up and smiled. 'Yer place is richt, Wull. Aye, richt, Son oan ye go,' and I followed my father down the road for my second shift.

'Ah, weel, here we are agin,' said Pop when we reached our roadhead after another journey mostly crouched double under the low roof. We stripped to singlet and trousers, and I could not help noticing how powerful he looked compared to me. He must have felt this, too, for he said 'Aye, ah'll no be long in buildin up some big muscles on ye. But sit doon an rest for a minute while ah check this roof. Ye'll be tired efter that walk.' He ducked under the roadhead, and I could hear him tapping his way along the roof with the head of his pick, listening carefully to the various sounds which told him so much about the condition of the rock above his workplace. There were broken props here and there, and I was surprised to see them, for we had left all the props well in place the previous day. The roof had been working, and the terrible pressures had splintered some of the props. 'Could ye saw some trees an put them up alongside they broken yins?' I was glad to try and eager to prove I could do another job efficiently. I measured the height with my pick and a piece of wood, sawed off the tree (as we sometimes called props) and erected it, tightening it into place with a short strap across the top, and hammered into position, making sure it was straight, so that the pressure came down true on it.

Pop came over and inspected my first prop. 'Aye,' he said, 'that's fine. Noo always mind an mak sure yer trees are up tight. Always watch yer heid. Just carry oan noo and replace all they broken trees, while ah try an get some coal for ye to shovel.' He went under the head, and I could hear his pick as he began, it seemed tirelessly, to cut under the coal.

As I worked away at replacing the broken props, there was the rattle of iron wheels on the rails and the jingle of the harness as Davie brought up the first hutch of the day. He greeted me, and I left my work to have a word with Sharp, who stood pat awaiting the next command. He nuzzled at my bare chest, and I gave him a piece of bread from my tin. He seemed grateful for it, and I stood a moment rubbing at his soft muzzle. He gave a little snort of pleasure, and stamped with his back hoof. 'Doan't ye be spoiling ma pownie,' Davie called, but I knew well enough that he did not mind anybody paying attention to his companion.

'How's it goin, Wull,' Davie called to my father. 'Auch, there's been a bit o weight on the place last nicht,' Pop replied, 'there's a lot o wood brokken.' Davie took a pick and went under the place to help my father for a while, and I took up the shovel to turn back the coal they were winning. Then I began filling the hutch, while Davie and Pop turned out the coal to the roadhead. Davie left with the full hutch, and promised that he would be back whenever he could to help with the pick work. I took up the pick he had been using and went under the coal myself, lying on my side, picking away at the hard surface, and turning back what I had loosened.

I was well under the head coal when Pop came across and warned me that it was time I had some stells up. 'Ye dinna want that lot to come doon on top of ye, he said, and helped me to put up the stells. He looked at the place, tapped it, and told me to hit the coal hard with the pick 'till it speaks back to ye'. This was an expression used when the weight of the head coal caused the bottom coal to burst out in flying fragments when struck by the pick. This was a dangerous time, and you had to be careful about your eyes, for those fragments burst out with a lot of force. The bottom coal I was howking was two or three inches thick, and then there was a layer of perhaps a foot of splint, a greyish, almost stone-like coal, and above that again was the head coal. We worked away, burrowing into the seam, lying on our sides by the flickering light of our lamps, our very lives depending on the stells we set. There was my father, an old miner, and his son, me, a lad of fourteen, deep under the earth, digging away at the very foundations of the world, winning coal, earning money literally by the sweat of our brows and the muscles of our bodies. It was a man's life.

B.L. Coombes
My First Night Underground

About nine-thirty that night I started to dress for my first night underground. There are no rules as to what you shall wear, only an unwritten one that you must not bring good clothes unless you do not mind being teased about what you are going to do for Sunday or 'how's it looking for the old 'uns?' Clothes must be tough and not too tight; dirtiness is no bar, because they will soon be much dirtier than they have ever been before. The usual wear is a cloth cap, old scarf, worn jacket and waistcoat, old stockings, flannel shirt, singlet, and pants. Thick moleskin trousers must be worn to bear the strain of kneeling and dragging along the ground, and strong boots are needed because of the sharp stones in the roadways and the other stones that fall. Food must be protected by a tin box, for the rats are hungry and daring; also plenty of tea or water is necessary to replace the sweat that is lost.

John was waiting for me near the colliery screens at ten o'clock. I was silent, and on edge for what the night would teach me. We climbed uphill between a double set of narrow rails, and the woods shut us in on either side. It took us half an hour of steady climbing before we halted on top of a grey pile that was the rubbish-tip and we could look back on the lights in the village below. The fire from the steelworks was like a red moonlight that night; we could see the expression on one another's faces by it. I was very near to my light, but did not get much comfort from that thought.

There were about fifty men sitting in the dark near the mouth of the level. We went inside a rough cabin that had closely printed extracts from the Mines Act and the Explosives Order nailed near the door. A youngish man was sitting on a box inside. I noticed that he was holding some pencilled notes to the left side of his face so that he could read them, and found out later that a piece of stone had knocked out his right eye some months before. He was the fireman, in charge of that shift because the colliery was not large enough to employ a night overman.

He seemed decent and intelligent. We handed in the 'starting-slip' to explain my presence, and he told John that he had found a strong-looking mate, and told me that I had got a good butty — one of the best in the valley. I am convinced the latter statement was correct.

At five minutes to eleven we started to walk inside. Each of us carried a 'naked light' or oil lamp that had no glass or gauze to restrict the light, because there was no danger of gas in that level.

As soon as we entered under the mountain I was aware of the damp atmosphere. Black, oily water was flowing continually along the roadway and out to the tip. It was up to the height of a man's knees, and to avoid it we had to balance carefully and talk along the narrow rails. I slipped several times, and then tried crouching up on the side and swinging myself along by the timber that was placed upright on either side. Suddenly I remembered that this timber was supposed to be holding the roof up and that I might pull it out of place and bring the mountain down on to us. I did not touch the timber after that. We were not more than ten minutes reaching the coal-face — that is the name given to the exact part where coal is being cut. This was a new level, so it had not gone far into the mountain. This seam was a small one, not a yard thick, and was a mixture of steam- and house-coal.

When I had been shown where to hang my clothes I went to see our working place. It was known as the Deep. We were the lowest place of all, because this Deep was heading into the virgin coal to open work. Every fifty yards on each side of our heading other headings opened left and right, but they would be working across the slope, and so were running level. From these level headings the stalls were opening to work all the coal off.

Our place was going continually downhill. Every three yards forward took us downward another yard. It was heavy climbing to go back, and every shovelful of coal or stone had to be thrown uphill. Water was running down the roadway to us and an electric pump was gurgling away on our right side. We were always working in about six inches of water, and if the pump stopped or choked for ten minutes the coal was covered with water. There is nothing pleasant about water underground. It looks so black and sinister. It makes every move uncomfortable and every stroke with the mandril splashes the water about your body.

It takes some time to be able to tell coal from the stone that is in layers above and below it. Everything is black, only the coal is a more shining black and the stone is greyer. It is difficult to tell one from the other, especially when water is about, but the penalty for putting stone — miners call it 'muck' —into a coal tram is severe. I tried very hard to be useful that night but was not successful, nor do I believe that any beginner ever has been. Things are so different and there is so much to learn. For several weeks lads of nowhere near my size and strength could make me look foolish when it came to doing the work they had been brought up in. I had used a shovel before, but found that skill was needed to force its round nose under a pile of rough stones on the uneven bottom, turn in that narrow space, and throw the shovelful some distance and to the exact inch. The need to watch where you step, the difficulty of breathing in the confined space, the necessity to watch how high you move your head, and the trouble of seeing under these strange conditions are all confusing until one has learned to do them automatically. It takes a while to learn that you must first take a light to a thing before you can find it. I started several times to fetch tools, then found myself in the solid darkness and had to return to get my lamp.

My mate lay on his side and cut under the coal. It took me weeks to learn the way of swinging elbows and twisting wrists without moving my shoulders. This holing under the coal was deadly monotonous work. We — or rather my mate — had to chip the solid coal away fraction by fraction until we had a groove under it of an inch, then six inches, then a foot. Then we threw water in the groove and moved along to a fresh place while the water softened where we had worked.

John hammered continually for nearly three hours at the bottom of the coal. He cut under it until he was reaching the full length of his arms and the pick-handle. At last he slid back and sat on his heels while he sounded the front of the coal with the mandril blade and looked closely at where the coal touched the roof to see if there showed the least sign of a parting.

'Keep away from this slip,' he warned me as he moved farther along, 'it'll be falling just now.'

It did, in less than five minutes, and after I had recovered from my alarm and most of the dust had passed I did my best to throw the coal into the tram. I soon found that a different kind of strength was needed than the one I had developed. My legs became cramped, my arms ached, and the back of my hands had the skin rubbed off by pressing my knee against them to force the shovel under the coal. The dust compelled me to cough and sneeze, while it collected inside my eyes and made them burn and feel sore. My skin was smarting because of the dust and flying bits of coal. The end of that eight hours was very soon my fondest wish.

After working for a while John went away to search for a post. About that time one of the hauliers — I never heard him called anything but Will Nosey — decided to go and see how this new starter was getting on. I was alone, and pretty nervous, when he arrived. His nickname came partly from his interest in the concerns of other people and partly because of his long nose, which curved downward as if it meant to get inside his mouth.

Our lighting was feeble, and his face showed a grey white in colour. His voice was always loud and the hollow passage amplified it. He had a different type of lamp from mine. His was one with an open wick and it was worn on his cap. The ventilation was good, so that the flame blew back over his head. I had reason to be alarmed, for his eyes were sunken alongside that hooked nose and there was a queer grin on his grey features, while above his head the oil flame hissed and crackled. He was exactly like one of those pictures we see of the Devil.

I did not speak, but stood and stared. He did exactly the same thing. Then someone not far away fired a shot to blow the roof down. In that confined space the noise and vibration were terrific. The whole mountain seemed to shake with the powder-charge. I could not hear for some seconds after. I had never before heard such a crash, nor had I had any warning to prepare me. I could only imagine that an explosion must have happened and that this other being with flame over his head was there to capture me.

Then the haulier, who had finished his scrutiny and had been delighted to see how startled I had been started to laugh. It was intended for laughter, but it sounded most sinister to me.

'He, he,' he chuckled, and when I realised that he used earthly words I became more at ease. 'He, he. Made you jump, did it? That was a shot for ripping top, that was. Oughter have warned us, they did; but I s'pose they was on the watch. Don't half shake things up, do it? Like to be here, hey?'

I had still enough determination left to say the lie that I did like being there. I was discovering still another new pain. My knees, unused to the hard rubbing against the stone bottom, had become like those of a Yorkshire miner whom I met later and who insisted that his knees had become 'b— red hot'.

As we were anxious to open work, we only cut the width of the roadway, about nine feet. The top coal was about one foot thick. We holed that off; then wedged up the lower coal. Altogether we filled three trams of coal, and the heading had gone forward about one yard. Then John rammed a bore-hole with powder and they lit the fuse. We went away to have food while the smoke from the shot was clearing.

We had a quarter of an hour for food. For the first time that I could remember I had no appetite, and the rats that ran about outside the circle of our lights had my food and squealed a lot while eating it.

The shot had ripped the top down, and we had to clear the stones so that the tram could be brought closer. Before that shot was fired we did not have more than a yard of height at the coal-face, and as I was clumsy I rubbed my back against the slime of the roof. My shirt-back was soon covered with a thick coat of clay and my back was getting as sore as my knees.

By four o'clock in the morning the shovel felt to be quite a hundredweight and I winced every time I touched my knees or back against anything. I got sleepy too, and felt myself swaying forward on my feet. I dropped some water on my eyes and revived for awhile, then I pinched my finger between two stones and was wide awake for some time after.

I had thought that night and day were alike underground, but it was not so. It is always dark, but Nature cannot be deceived, and when the time is night man craves for sleep. When the morning comes to the outside world he revives again, as I did.

Even the earth sleeps in the night and wakens with perceptible movement about two o'clock in the morning. With its waking shudders it dislodges all stones that are loose in the workings. It is about that time that most falls occur, at the time when man's energy is at its lowest.

Somehow that shift did end, although I felt it lasted the time of two. Then we had to lock our tools for the day. Holes are bored in the handles of the tools and they are pushed on a thin steel bar with a locking-clip fitted in the end.

John had a pile of tools, and they were all needed. Shovels, mandrils of different sizes, prising-bars; hatchet, powder-tin and coal-boxes, boring-machine and drills and several other things. He valued them at eight pounds worth, and he was forced to buy them himself. He knew they might be buried by a fall any day and was not hopeful of getting any compensation for them. Nearly every week he had to buy a new handle of some sort and fit it into the tool at his home, so that his wages were not all clear benefit, and his work not always finished when he left the colliery.

How glad I was to drag my aching body toward that circle of daylight! I had sore knees and was wet from the waist down. The back of my right hand was raw and my back felt the same. My eyes were half-closed because of the dust and my head was aching where I had hit it against the top but I had been eight hours in a strange, new world.

The outside world had slept while we worked, and the dew of the morning sparkled from a thousand leaves when I looked down on the valley. It was beautiful after that wet blackness to see the sun, and the brown mountain, and the picture that the church tower made peeping out over the trees.

As we went less down the incline, the day-shift came up. They called their 'Good morning' or 'Shu mai?' as they hurried past with their tea-jacks in their hands.

Vernon Watkins
The Collier

When I was born on Amman hill
A dark bird crossed the sun.
Sharp on the floor the shadow fell;
I was the youngest son.

And when I went to County School
I worked in a shaft of light.
In the wood of the desk I cut my name:
Dai for Dynamite.

The tall black hills my brothers stood;
Their lessons all were done.
From the door of the school when I ran out
They frowned to watch me run.

The slow grey bells they rang a chime
Surly with grief or age.
Clever or clumsy, lad or lout,
All would look for a wage.

I learnt the valley flowers' names
And the rough bark knew my knees.
I brought home trout from the river
And spotted eggs from the trees.

A coloured coat I was given to wear
Where the lights of the rough land shone.
Still jealous of my favour
The tall black hills looked on.

They dipped my coat in the blood of a kid
And they cast me down a pit,
And although I crossed with strangers
There was no way up from it.

Soon as I went from the County School
I worked in a shaft. Said Jim,
'You will get your chain of gold, my lad,
But not for a likely time.'

And one said, 'Jack was not raised up
When the wind blew out the light
Though he interpreted their dreams
And guessed their fears by night.'

And Tom, he shivered his leper's lamp
For the stain that round him grew;
And I heard mouths pray in the after-damp
When the picks would not break through.

They changed words there in darkness
And still through my head they run,
And white on my limbs is the linen sheet
And gold on my neck the sun.

The Working Day

Anon.
The Collier's Rant

As me and my marrow was ganning to wark,
We met with the devil, it was in the dark;
I up with my pick, it being in the neit,
I knock'd off his horns, likewise his club feet
    Follow the horses, Johnny my lad Oh!
    Follow them through, my canny lad Oh!
    Follow the horses, Johnny my lad Oh!
    Oh lad ly away, canny lad Oh!

As me and my marrow was putting the tram,
The lowe it went out, and my marrow went wrang;
You would have laugh'd had you seen the gam,
The deil gat my marrow, but I gat the tram,
    Follow the horses, etc.

Oh! marrow, Oh! marrow, what dost thou think?
I've broken my bottle, and spilt a' my drink;
I lost a' my shin-splints among the great stanes,
Draw me t' the shaft, it's time to gan hame
    Follow the horses, etc.

Oh! marrow, Oh! marrow, where hast thou been?
Driving the drift from the low seam,
Driving the drift from the low seam;
Had up the lowe, lad, deil stop out thy een!
    Follow the horses, etc.

Oh! marrow, Oh! marrow, this is wor pay week,
We'll get penny loaves and drink to our beek;
And we'll fill up our bumper and round it shall go,
Follow the horses, Johnny lad Oh!
    Follow the horses, etc.

There is my horse, and there is my tram;
Twee horns full of greese will make her to gang;
There is my hoggars, likewise my half shoon,
And smash my heart, marrow, my putting's a' done.
    Follow the horses, Johnny my lad Oh!
    Follow them through, my canny lad Oh!
    Follow the horses, Johnny my lad Oh!
    Oh lad ly away, canny lad Oh!

George Orwell
Different Universes

When you have been down two or three pits you begin to get some
grasp of the processes that are going on underground. (I ought to
say, by the way, that I know nothing whatever about the technical
side of mining: I am merely describing what I have seen.) Coal
lies in thin seams between enormous layers of rock, so that
essentially the process of getting it out is like scooping the central
layer from a Neapolitan ice. In the old days the miners used to cut
straight into the coal with pick and crowbar — a very slow job
because coal, when lying in its virgin state, is almost as hard as
rock. Nowadays the preliminary work is done by an electrically-
driven coal-cutter, which in principle is an immensely tough and
powerful band-saw, running horizontally instead of vertically,
with teeth a couple of inches long and half an inch or an inch thick.
It can move backwards or forwards on its own power, and the men
operating it can rotate it this way and that. Incidentally it makes
one of the most awful noises I have ever heard, and sends forth
clouds of coal dust which make it impossible to see more than two
or three feet and almost impossible to breathe. The machine
travels along the coal face cutting into the base of the coal and
undermining it to the depth of five feet or five feet and a half; after
this it is comparatively easy to extract the coal to the depth to
which it has been undermined. Where it is 'difficult getting',
however, it has also to be loosened with explosives. A man with
an electric drill, like a rather smaller version of the drills used
street-mending, bores holes at intervals in the coal, inserts blasting
powder, plugs it with clay, goes round the corner if there is one
handy (he is supposed to retire to twenty-five yards distance) and
touches off the charge with an electric current. This is not intended to bring the coal out, only to loosen it. Occasionally, of
course, the charge is too powerful, and then it not only brings the
coal out but brings the roof down as well.

After the blasting has been done the 'fillers' can tumble the coal
out, break it up, and shovel it on to the conveyor belt. It comes
out at first in monstrous boulders which may weigh anything up
to twenty tons. The conveyor belt shoots it on to tubs, and the
tubs are shoved into the main road and hitched on to an endlessly
revolving steel cable which drags them to the cage. Then they are
hoisted, and at the surface the coal is sorted by being run over
screens, and if necessary is washed as well. As far as possible the
'dirt' — the shale, that is — is used for making the roads below.

All that cannot be used is sent to the surface and dumped; hence the monstrous 'dirt-heaps', like hideous grey mountains, which are the characteristic scenery of the coal areas. When the coal has been extracted to the depth to which the machine has cut, the coal face has advanced by five feet. Fresh props are put in to hold up the newly exposed roof, and during the next shift the conveyor belt is taken to pieces, moved five feet forward, and re-assembled. As far as possible the three operations of cutting, blasting, and extraction are done in three separate shifts, the cutting in the afternoon, the blasting at night (there is a law, not always kept, that forbids its being done when there are other men working near by), and the 'filling' in the morning shift, which lasts from six in the morning until half past one.

Even when you watch the process of coal-extraction you probably only watch it for a short time, and it is not until you begin making a few calculations that you realize what a stupendous task the 'fillers' are performing. Normally each man has to clear a space four or five yards wide. The cutter has undermined the coal to the depth of five feet, so that if the seam of coal is three or four feet high, each man has to cut out, break up, and load on to the belt something between seven and twelve cubic yards of coal. This is to say, taking a cubic yard as weighing twenty-seven hundredweight, that each man is shifting coal at a speed approaching two tons an hour. I have just enough experience of pick and shovel work to be able to grasp what this means. When I am digging trenches in my garden, if I shift two tons of earth during the afternoon, I feel that I have earned my tea. But earth is tractable stuff compared with coal, and I don't have to work kneeling down, a thousand feet underground, in suffocating heat and swallowing coal dust with every breath I take; nor do I have to walk a mile bent double before I begin. The miner's job would be as much beyond my power as it would be to perform on the flying trapeze or to win the Grand National. I am not a manual labourer and please God I never shall be one, but there are some kinds of manual work that I could do if I had to. At a pinch I could be a tolerable road-sweeper or an inefficient gardener or even a tenth-rate farm hand. But by no conceivable amount of effort or training could I become a coal-miner; the work would kill me in a few weeks.

Watching coal-miners at work, you realize momentarily what different universes different people inhabit. Down there where coal is dug it is a sort of world apart which one can quite easily go through life without ever hearing about. Probably a majority of people would even prefer not to hear about it. Yet it is the absolutely necessary counterpart of our world above. Practically everything we do, from eating an ice to crossing the Atlantic, and from baking a loaf to writing a novel, involves the use of coal, directly or indirectly. For all the arts of peace coal is needed; if war breaks out it is needed all the more. In time of revolution the miner must go on working or the revolution must stop, for revolution as much as reaction needs coal. Whatever may be happening on the surface, the hacking and shovelling have got to continue without a pause, or at any rate without pausing for more than a few weeks at the most. In order that Hitler may march the goosestep, that the Pope may denounce Bolshevism, that the cricket crowds may assemble at Lord's, that the Nancy poets may scratch one another's backs, coal has got to be forthcoming. But on the whole we are not aware of it; we all know that we 'must have coal', but we seldom or never remember what coal getting involves. Here am I, sitting writing in front of my comfortable coal fire. It is April but I still need a fire. Once a fortnight the coal cart drives up to the door and men in leather jerkins carry the coal indoors in stout sacks smelling of tar and shoot it clanking into the coal-hole under the stairs. It is only very rarely, when I make a definite mental effort, that I connect this coal with that far-off labour in the mines. It is just 'coal ' — something I have got to have; black stuff that arrives mysteriously from nowhere in particular, like manna except that you have to pay for it. You could quite easily drive a car right across the north of England and never once remember that hundreds of feet below the road you are on the miners are hacking at the coal. Yet in a sense it is the miners who are driving your car forward. Their lamp-lit world down there is as necessary to the daylight world above as the root is to the flower.

It is not long since conditions in the mines were worse than they are now. There are still living a few very old women who in their youth have worked underground,with a harness round their waists and a chain that passed between their legs, crawling on all fours and dragging tubs of coal. They used to go on doing this even when they were pregnant. And even now, if coal could not be produced without pregnant women dragging it to and fro, I fancy we should let them do it rather than deprive ourselves of coal. But most of the time, of course, we should prefer to forget that they were doing it. It is so with all types of manual work; it keeps us alive, and we are oblivious of its existence. More than anyone else, perhaps, the miner can stand as the type of the manual worker, not only because his work is so exaggeratedly awful, but also because it is so vitally necessary and yet so remote from our experience, so invisible, as it were, that we are capable of forgetting it as we forget the blood in our veins. In a way it is even humiliating to watch coal-miners working. It raises in you a momentary doubt about your own status as an 'intellectual' and a superior person generally. For it is brought home to you, at least while you are watching, that it is only because miners sweat their guts out that superior persons can remain superior. You and I and the editor of the *Times Lit. Supp.*, and the Nancy poets and the Archbishop of Canterbury and Comrade X, author of *Marxism for Infants* — all of us *really* owe the comparative decency of our lives to poor drudges underground, blackened to the eyes, with their throats full of coal dust, driving their shovels forward with arms and belly muscles of steel.

D.H. Lawrence
Fuses

The only times when he entered again into the life of his own people was when he worked, and was happy at work. Sometimes, in the evening, he cobbled the boots or mended the kettle or his pit-bottle. Then he always wanted several attendants, and the children enjoyed it. They united with him in the work, in the actual doing of something, when he was his real self again.

He was a good workman, dextrous, and one who, when he was in a good humour, always sang. He had whole periods, months, almost years, of friction and nasty temper. Then sometimes he was jolly again. It was nice to see him run with a piece of red-hot iron into the scullery, crying:

'Out of my road — out of my road!'

Then he hammered the soft, red-glowing stuff on his iron goose, and made the shape he wanted. Or he sat absorbed for a moment, soldering. Then the children watched with joy as the metal sank suddenly molten, and was shoved about against the nose of the soldering-iron, while the room was full of a scent of burnt resin and hot tin, and Morel was silent and intent for a minute. He always sang when he mended boots because of the jolly sound of hammering. And he was rather happy when he sat putting great patches on his moleskin pit trousers, which he would often do, considering them too dirty, and the stuff too hard, for his wife to mend.

But the best time for the young children was when he made fuses. Morel fetched a sheaf of long sound wheat-straws from the attic. These he cleaned with his hand, till each one gleamed like a stalk of gold, after which he cut the straws into lengths of about six inches, leaving, if he could, a notch at the bottom of each piece. He always had a beautifully sharp knife that could cut a straw clean without hurting it. Then he set in the middle of the table a heap of gunpowder, a little pile of black grains upon the white-scrubbed board. He made and trimmed the straws while Paul and Annie filled and plugged them. Paul loved to see the black grains trickle down a crack in his palm into the mouth of the straw, peppering jollily downwards till the straw was full. Then he bunged up the mouth with a bit of soap — which he got in his thumb-nail from a pat in a saucer — and the straw was finished.

'Look, dad!' he said.

'That's right, my beauty,' replied Morel, who was peculiarly lavish of endearments to his second son. Paul popped the fuse into the powder-tin, ready for the morning, when Morel would take it to the pit, and use it to fire a shot what would blast the coal down.

Meantime Arthur, still fond of his father, would lean on the arm of Morel's chair, and say:

'Tell us about down pit, daddy.'

This Morel loved to do.

'Well, there's one little 'oss — we call 'im Taffy,' he would begin. 'An he's a fawce un!'

Morel had a warm way of telling a story. He made one feel Taffy's cunning.

'He's a brown un,' he would answer, 'an' not very high. Well, he comes i' th' stall wi' a rattle, an' then yo' 'ear 'im sneeze. "'Ello, Taff," you say, "what art sneezin' for? Bin ta'ein' some snuff?"

'An' 'e sneezes again. Then he slives up an' shoves 'is 'ead on yer, that cadin'.

' "What's want, Taff?" yo' say!'

'And what does he?' Arthur always added.

'He wants a bit o' bacca, my duckey.'

This story of Taffy would go on interminably, and everybody loved it.

Or sometimes it was a new tale.

'An' what dost think, my darlin'? When I went to put my coat on at snap-time, what should go running up my arm but a mouse.

'"Hey up, theer!" I shouts.

'An' I wor just in time ter get 'im by th' tail.'

'And did you kill it?'

'I did, for they're a nuisance. The place is fair snied wi' 'em.'
'An' what do they live on?'

'The corn as the 'osses drops — and they'll get in your pocket an' eat your snap, if you'll let 'em — where yo' hing your coat — the slivin', nibblin' little nuisances, for they are.'

These happy evenings could not take place unless Morel had some job to do. And then he always went to bed very early, often before the children. There was nothing remaining for him to stay up for, when he had finished tinkering, and had skimmed the headlines of the newspaper.

And the children felt secure when their father was in bed. They lay and talked softly awhile. Then they started as the lights went suddenly sprawling over the ceiling from the lamps that swung in the hands of the colliers tramping by outside, going to take the nine o'clock shift. They listened to the voices of the men, imagined them dipping down into the dark valley. Sometimes they went to the window and watched the three or four lamps growing tinier and tinier, swaying down the fields in the darkness. Then it was a joy to rush back to bed and cuddle closely in the warmth.

Sid Chaplin
The Shaft

Williamson looked up from the bunch of reports he was reading. Although he smiled around the lips, his eyes were worried.

'Sit down, Harry,' he said. 'Ah'll get through these first.' Harry sat down uneasily, rumpling his cap. Williamson went on with his pretence of reading the reports. There came another knock at the office door. Both men turned. Williamson put the reports carefully on to the desk. 'Come in !' he called out.

The man who entered could have been Harry's double. There was the same solid, medium build and sandy hair. Blue eyes set in a square face. A hard, uncompromising mouth. His eyes darted from Williamson to Harry, then back to Williamson. 'Morning, Mr Williamson,' he said. There was a strain of reproach mingling with the respect in his voice.' 'Good morning, Willy,' Williamson said. 'Find a seat.' Willy found a seat, as far away from Harry as was possible.

Harry stood up. His lips were one thin bitter line. 'Ah dare say our business can wait, Mr Williamson. Ah'll be at hand when ye want me.' He made for the door. Williamson sighed. This was going to be a tough job. 'You can sit down again, Harry. Ah want a word with both of ye.' He leaned back in his chair and tried to assume confidence.

'You'll both be wondering why Ah've asked you to call here to-day, eh?' Both men shifted uncomfortably. 'Well, it's awkward, dam' awkward, but it's got to stop, see? Seems everybody knew about this but me, or you'd have been on the carpet before now. What you do outside of work isn't any concern of mine, but when two men on a job like yours aren't on speaking terms, it's time to draw the line.' His eyes roved from the ceiling to the two silent men. Both stared back with rising hostility.

'What's the grumble?' asked Willy. 'We're keeping the shafts good; we do our work proper; so with all due respect to you, Mr Williamson, Ah think you're out of order.'

Williamson waved a placatory hand. 'No grumbles about your work; none, none at all. But this is what Ah'm getting at. You two don't speak? Right? Well, it's none of my business why, but remember you're on a responsible job. And accidents can happen in them shafts. Ah've seen one or two in my time. Accidents, mind ye. And what's people going to say if anything does happen to one of ye, eh?'

There was a long silence. 'We know our work,' said Harry at last. 'Nothing's going to happen.'

'It's between him and me,' said Willy. 'Personal; nothing to do with work, or with you.'

'That's right,' said Harry. 'It's our affair and ye needn't worry about it interfering with our work.'

Williamson thumped the table. 'My God!' he cried. 'Are ye men, or just bits of bairns? Can't ye see what Ah'm getting at? With the best will in the world anything can happen in a pit shaft....' As he spoke the words a sudden terrifying vision flashed across his mind. He saw two hundred fathoms of shaft vertically piercing the strata; a shadowy hole diminishing into extreme blackness. He saw the pipe-lined sides; pipes sucking unending gallons of water from underground dams; pipes carrying electric cables, and pipes through which rushed air, compressed to drive a thousand drills through virgin coal and naked rock. He saw the greasy black cables of the guides quiver like the key-cable of a web when a fly is enmeshed, and then a cage flash by with two figures on top, two men holding to the guy chains almost nonchalantly.

He switched his mind to the office; he had seen that picture a few times in the last few days; and always with a singularly sinister ending. Williamson possessed an imagination.

He raked both men with a significant glance. 'So it's got to stop! You can cut each other dead in the street, but by God, you'll speak at work! Either that, or one of you goes. That's final.' He fumbled with his reports. Both men remained seated, dumbly staring at him. 'That's all,' he said.

They walked out. Through his office window he saw them turn in opposite direction. A pity, he thought, two good workmen and brothers at that. Brothers! Well, they're not doing a Cain and Abel on this job. Give them a week, no more.' Then one of them goes. Or both for that matter. That one or the other might capitulate never entered his mind. Williamson knew his men.

When Harry got home he found his wife busy with rolling pin and dough, baking fadges and red with the fire's heat. A neat little woman with sharp features enclosed by a mop of hair which had earned for her the nick-name of Ma Golliwog among the local bairns. 'Well?' she asked sharply. 'And what did Mr Williamson want with ye?'

He settled down gloomily in the big rocking-chair. 'Come to his ears about me an' Willy never speaking. Says it won't do.' She slammed the rolling pin on to the table. 'Well, Ah'll be jiggered! A fine thing it is when the bosses start interfering with the men's family affairs. It won't do, eh? Well, it'll have to do, 'cos Ah'll never speak to that lot again! And Mr Williamson can like it or lump it.'

She picked up her rolling pin and started to stamp, rather than roll, the unhappy dough. Then she looked up suspiciously. And what had you to say to him ?'

'Told him it was our affair, and so did Willy.'

'Willy? Ye don't mean to tell me that he had ye both there? The impudence of the man!' she screeched, her eyes flashing fire.

'Said that if we don't settle the thing he's going to sack one of us.'

'Well, now, Harry Ward,' she said, quiet now, 'if you start speaking to that brother of yours, Ah'll leave you. So that's that.'

'Catch me speaking to him,' he said. 'Ah'll not forget him in a hurry.' But as he rocked the old chair he got to wondering how that slight seven stone of woman could hold so much bitterness. He sighed, then lit his pipe.

At the other side of the village much the same scene occurred at the home of Willy Ward. Willy may have been a little gloomier and his wife a trifle more bitter. After hearing the story, she repeated her determination not to patch up the quarrel, come what might. 'But what if Ah lose me job, honey?' said Willy. 'Might easily be me that has to go.'

'It'll not be you,' said his wife, confidently. 'You're twice as good a man as him. So you're not going to be first to break silence. Not after the way they carried on the day your Dad was buried. And all because he left them pictures to you. Not that they're very good pictures, they're not, and if it wasn't for the principle of the thing Ah'd let them have them. But there it is; the will was read, it was there in black and white for all the world to see! And then that screaming vulgar Lizzie had to start shoutin' at the top of her voice that they'd been promised to her. The very idea! No, it was them that started it and they shall finish it. D'ye hear what Ah say, Willy Ward?'

'Aye,' answered her husband, and went to bed for some sleep and peace. When he awoke the room was dark. A rattling window told of a high wind outside. He lay a moment, wishing he had another job and could lie in bed o' nights, instead of going out to work when others were going to sleep. Then he remembered the interview with Williamson and realized that his wish might soon come true. Groaning, he got out of bed and made his way downstairs. His supper was laid ready for him. 'Bit of a wind up,' he remarked. 'Aye, it's blown tiles off every roof in the street. And Mrs Roberts come in to say that the roof's been blown off the tin chapel!'

He had his supper and set off for work after another warning that there was to be no pact 'unless he speaks first.'

It *was* a wind! It billowed his raincoat out like a balloon and blew him, almost carried him, to the pit. Several times tiles and slates whizzed past his head to splinter into a thousand pieces at his feet. He was glad when he got to work. Harry was already in their cabin, changing into the warm, heavy clothes and buckling on his safety straps and belt. On a slate was chalked their work for that night.' Loose guides in No. 1 shaft. Burst water-pipe in No. 2.' He glanced at the slate and proceeded to change.

It wasn't necessary to discuss which job should be done first.

The burst pipe took precedence, since it would take longer to repair. Willy changed while Harry collected the gear needed. Not a word was said. When they went out the wind was blowing a gale. Harry was first out and he was swept back into the cabin like a ninepin, almost knocking Willy on to his back. There was no apology given, or expected. They struggled against the wind until they reached 'the hole' at the base of the headstocks, then passed through an air-lock all entered the steel superstructure which covered the shaft. There they stood a moment, watching the cages sweep past with their cargoes of coal. Finally the electric bell above their heads shrilled six times. A door on the landing-level above them opened and the banksman appeared. 'All clear down there,' he shouted. Both men nodded. Used to their ways, he descended the spider-ladder to the duplicate controls. 'She's a raw, rough wind tonight,' he said. 'Aye she's rough,' said Harry. He tightened his safety-belt one hole. Worry had kept him from eating since the interview with Williamson. Davies rapped the cage to the level of the hole. It glided slowly into view, glistening wet with the sprayed water from the burst. The shaftmen stepped on the roof, fastened their safety-belts to the guy-chains, hung the gong, their only method of communication with Davies, then switched on their cap-lamps.

'Ready?' asked Davies. The wailing of the wind above, striking music from the pulley-wheels and girders, made him uneasy. There was something about the two men, also, that made for uneasiness, the way they talked with a flicker of an eyelid, or a motion of the hand. 'It's not natural,' he muttered. 'Brothers at that!' He watched intently, waiting for a nod from each. Each man took hold of the nearest guy-chain, then nodded. He rapped away and watched the cage slowly sink into the black hole until it disappeared. Then, attaching the chain of the manual-signal to his hand, he walked to the edge of the shaft and peered down. He saw the two figures on the cage-top, curiously foreshortened in the glow of their lamps, like men viewed through the wrong end of a telescope. Indeed, the shaft itself, with its diminishing wet walls, was like the inside of a gigantic telescope. The gong sounded. Its peal came echoing out of the depths, faint against the background of the gale that raged outside. He pulled the manual-chain urgently and sighed with relief as the greasy black steel ropes came to a standstill.

Down below, the cage was level with the burst pipe. A stream of water was sprayed from it, and the two men were soon soaked. They had a pair of clamps ready, but the pipe was out of reach when their safety-belts were attached, so they unfastened them, and proceeded to fix the clamps. Harry had the inner clamp, and this meant he had to lie on the cage-top, holding his clamp in position with one hand while he held on to a guy-chain with the other, and Willy attached the other clamp and bolted it. It was an uncomfortable job for Harry, since the dripping water was ice-cold and his arm ached with the weight of the clamp. And it was just at this moment that the gale above reached its peak, dislodging a rusting, insecure section of corrugated iron roofing. The wind lifted it as if it were no heavier than a sheet of newspaper, and pitched it unerringly into the narrow outlet of the superstructure. It swept past the astonished Davies and entered the shaft, ricocheting from side to side. The two men heard it coming. Harry pulled himself back on to his knees, letting the clamp drop. It crashed to the shaft-bottom, followed a split-second later by the section of roofing, which brushed, just brushed, Harry in passing.

But it was enough to knock him off his balance, and he would have followed clamp and roof-section had not Willy, by a purely reflex action, which was perfectly timed, caught him by the wrist. Harry's thirteen stone pulled Willy to the roof of the cage, but he still managed to hold on, with both hands clamped around one of Harry's wrists. He tried to pull his brother back, but found he was in no position for lifting; indeed, he imagined he heard his arms scrape in their sockets with the downward pull. Harry's dead-white face gazed up at him, running with sweat and water mixed. For the burst pipe was spraying directly above their heads. And, kick as hard as he could, Harry could not find even a bolt-head to support himself upon.

There was only one thing to do and that was to get the cage back to the level where Davies could help. But the gong was well out of reach. He locked one of his feet round a guy-chain and tried to kick out at the gong with the other. But it was too far away. There was only one thing left to do. He shouted. The Bull of Bashan had nothing on Willy that night. His desperate yell floated out of the blackness, and Davies, leaning over the shaft edge, heard it above the gale. '*Raise her to bank, Joe!*'

Davies pulled the manual chain and watched the sliding ropes, with his heart pounding. When the cage came into view he saw the situation immediately. He rapped hold and leaped on to grab the other arm. Even so, the two of them had a struggle to haul Harry back to safety. For a full five minutes all three lay on the cage top, panting for breath. Then Willy stood up. The other brother tried to follow suit but failed.

'Take it easy, now,' said Willy. Harry looked up. He had not heard that voice for months; he had not heard that particular note in it since they were lads together, the day he'd fallen in the river and lost a stocking, and Willy had met him going home, crying his eyes out. Then he smiled slowly. 'Always knew you'd be the one to break t' silence,' he said.

Willy had no immediate answer for this. His mind was on those few minutes of agony in the shaft. He had seen one man fall the full two hundred fathoms; he had seen the shattered, shapeless body wrapped in canvas. The Missus can play hell about me speaking first, he thought. Imagine Harry, me brother, lying down there, dead and broken. But he's alive, and Ah'd suffer a thousand nagging women for the joy of it.

'Mebbe Ah did break silence,' he said, 'but somebody had to speak, and since thou was so shook wi' fear, and Joe was out of breath a bit, it had to be me.'

Anon.
The Colliers' Pay Week

The Baff Week is o'er — no repining —
Pay-Saturday's swift on the wing;
At length the blithe morning comes shining,
When kelter makes colliers sing.
'Tis spring, and the weather is cheery,
The birds whistle sweet on the spray;
Now coal-working lads, trim and airy,
To Newcastle town hie away.

Those married jog on with their hinnies,
Their canny bairns go by their side;
The daughters keep teasing their minnies
For new clothes to keep up their pride.
They plead Easter Sunday does fear them,
For if they have nothing that's new,
The crow, spiteful bird, will besmear them.
Oh, then what a sight for to view!

The young men, full blithesome and jolly,
March forward, all decently clad;
Some lilting up *Cut-and-Dry Dolly*,
Some singing *The Bonny Pit Lad*.
The pranks that were played at last binding
Engage some in humorous chat.
Some halt by the wayside on finding
Primroses to place in their hat.

Bob Cranky, Jack Hogg and Dick Marley,
Bill Hewitt, Luke Carr and Tom Brown,
In one jolly squad set off early
From Benwell to Newcastle Town.

Such hewers as they (none need count it)
Ne'er handled a shovel or pick.
In high or low seam they could suit it,
In regions next door to Old Nick.

Some went to buy hats and new jackets,
And others to see a bit fun;
And some wanted leather and tackets
To cobble their canny pit shoon.
Save the ribbon Dick's dear had requested
— Aware he had plenty of chink —
There was no other care him infested,
Unless 'twere his care for good drink.

At length in Newcastle they centre,
In Hardy's, a place much renowned,
The jovial company enter
Where stores of good liquor abound.
As quick as the servants could fill it
— Till emptied were quarts half a score —
With heart-burning thirst down they swill it,
And thump on the table for more.

With boozing and laughing and smoking,
The time slippeth swiftly away;
And while they are ranting and joking,
The church clock proclaims it mid-day.
And now for black puddings, long measure,
They go to Tib Trollibag's stand,
And away bear the glossy rich treasure
With joy, like curled bugles in hand.

And now a choice home they agreed on,
Not far from the head of the Quay,
Where they their black puddings might feed on,
And spend the remains of the day,
Where pipers and fiddlers resorted
To pick up the straggling pence,
And where the pit lads often sported
Their money at fiddle and dance.

Blind Willie the fiddler sat scraping
In a corner just as they went in.

Some Willington callants were shaking
Their feet to his musical din.
Jack vowed he would have some fine capering,
As soon as their dinner was o'er,
With the lassie that wore the white apron,
Now reeling about on the floor.

Their hungry stomachs being eased,
And gullets well cleared with a glass,
Jack rose from the table and seized
The hand of the frolicsome lass.
'Maw hinny,' says he, 'pray excuse me —
To ask thee to dance aw make free.'
She replied: 'Aw'd be loth to refuse thee.
'Now fiddler, play *Jigging for me.*'

The damsel displays all her graces,
The collier exerts all his power;
They caper in circling paces,
And set at each end of the floor.
He jumps, and his heels knack and rattle,
At turns of the music so sweet.
He makes such a thundering brattle,
The floor seems afraid of his feet.

This couple being seated, Bob rose up.
He wished to make one in a jig;
But a Willington lad set his gob up —
O'er him there should none 'run the rig'.
For now 'twas his turn for a caper,
And he would dance first as he'd rose.
Bob's passion began for to vapour;
He twisted his opponent's nose.

The Willington lads, for their Frankie,
Jumped up to revenge the foul deed;
And those in behalf of Bob Cranky
Sprung forward, for now there was need.
Bob canted the form with his kevel,
As he was exerting his strength;
But he got on the lug such a nevel
That down he came, all his long length.

All hand-over-hand, topsy-turvy,
They struck with fists, elbows and feet.
A Willington callant called Gurvy
Was top-tails tossed over the seat.
Luke Carr had one eye closed entire,
And what is a serio-farce,
Poor Robin was cast on the fire.
His breeks torn and burnt off his arse.

Oh, Robin, what argued thy speeches?
Disaster now makes thee quite mum.
Thy wit could not save the good breeches
That mensefully covered thy bum.
To some slop-shop now thou may go trudging,
And lug out some squandering coins;
For now 'tis too late to be grudging;
Thou cannot go home with bare groins.

How the wayfaring companies parted
The Muse chooseth not to proclaim;
But 'tis thought that, being rather downhearted,
They quietly went toddling hame.
Now, ye collier callants se clever,
Residing 'tween Tyne and the Wear,
Beware, when you fuddle together,
Of making too free with strong beer.

Jack Jones
Working a Stall

Joe's dad could, so Joe thought, work a 'stall' better than any man
in any pit anywhere. The way he used to work the coal forward,
holing with his holing-mandrel the face-slips to make the 'slips'
of coal give to make them fall. You had to 'hole' under the coal
of the stalls on the face-side of the heading, but on the other side,
which was called 'the back-side', you cleaned what was called the
'brug' in Welsh which in English no doubt meant the 'brow' of
the backward-sloping slips of coal. The slips like layers of coal
varied in thickness, and they were easier to work when thin — say
a foot thick — than when they were thick, say a yard thick. After
loosening coal and loading it into trams, the empty space left had

to be timbered to hold the roof up, and also filled with the rock blasted to make a roadway of six feet high to enable the tram to follow the coal forward. All sorts of jobs Joe's dad could do, and all of them well, and Joe was now learning to do many of the jobs that required doing. Sometimes — and it used to frighten Joe at first, but not now — when his father was 'holing butts' or 'cleaning brugs' to loosen the coal, the coal would 'pounce' angrily. A 'pounce' is a muffled little explosion which the coal makes in its anger when men all over the pit are worrying it and disturbing its rest. First, as it was being disturbed, it would start making complaints like firework crackers which grew, louder and louder until they would culminate in 'pounce', which was its surrender to the man attacking it so cunningly and ruthlessly. The slip of coal would groan as it cracked to fall on the rock bottom. In the dust it made falling Joe's father would roll over on his back and chuckle triumphantly. 'There it is, Joe bach. Now to fill it into the tram — but reach me my tea-jack first.' Steaming with sweat he would gargle his throat to remove the thickest of the coal-dust before drinking deep. 'Now a chew, an' then we'll start filling.' Out with his brass tobacco-box from his trousers' pocket. Loose tobacco, about half a pipeful, into his mouth, before he started handling the hugest lumps of coal in a way that never failed to amaze Joe. Oh, the crossbar lump with which he made secure the 'door-end' of the walled-up tram of coal, the end which was opened at the pit-head as the tram after it was weighed went forward to be tipped into railway trucks. Yes, my dad knows how to fill a tram of coal, Joe thought. Nearly thirty hundred-weight he could build like a house founded on rock into each tram.

H.V. Morton
Colliery Horses

On the way back from the 'front line' I asked to be shown the 'pit ponies'.

'There are no pit ponies in Wales,' I was told. 'They are colliery horses.'

I was taken to the mine stables near the pit bottom. There were six stalls well lit by electric light. A man was whitewashing them. The horses not on duty were being groomed. I noticed that above every stall was written the name of its occupant, as in racing stables. Each pit horse is christened before it goes down a mine.

No colliery horse is under 14.2 hands high. They cost between
£55 and £60 each. One colliery values its horses at over £7,000.
' "Warrior" has been underground for fifteen years,' said the
farrier. 'Does he look unhappy? Does he look ill-fed?'
He did not. 'How many of these horses are blind?'
'I have never met a blind pit horse,' said the farrier. 'It is true
that after a number of years a horse taken above has defective sight;
but they are never totally blind.
'They get the best of food, good quarters, and work which is not
so strenuous as that of a London dray horse ....'
'What about injuries?'
'After every shift the haulier is forced to report any injury no
matter how trivial. The slightest scratch must be reported. The
vet then visits the horse at once and treats him if necessary.
Cruelty? We never meet it. But there are very strict regulations in
every colliery to guard against it. If a haulier is cruel to his horse
a red mark goes down against his name. This is against him even
if his action might have been accidental. Three red marks and a
man is instantly sacked. There is no argument. Out he goes!'
The live-stock in a coal mine is interesting. A fat black cat was
sitting in these stables on a bale of hay. All mines have cats on the
ration list. They keep down the mice which get into the fodder.
Colliery books contain dozens of entries like this:

Milk . . . . 1s. 6d.

The haulier has charge also of the stable cat and its milk.
'Some of these cats are funny,' said the farrier. 'They will go to
the pit bottom and wait for the cage like a man. They walk in, go
up to the surface, take a look round to see if it's raining and wait
for the next cage down again! Sometimes the hauliers take them
home for a week-end.'
We watched the hauliers bringing their horses to stables after
the shift. The pit horse has a journey of perhaps 200 yards from
the coal face to the junction. He draws 28 cwt. of coal in a 'tram'
on rails. He works six or eight shifts a week, but he is never allowed
to work two consecutive shifts. He never works on Saturday or
Sunday.
People who think that a pit horse just pulls a 'tram ' have never
seen these animals at work. The horse, like the haulier, knows his
job. Every haulier keeps to his own horse, and would very much
resent any other man working him. And the reason is that in taking
coal from the seam to the junction there are various little difficul-
ties which can only be easily avoided by a sympathy between man
and beast.

When, for instance, a 'tram' runs off the line or fouls the points, a man and horse who know each other's methods of work can rail it again in a few seconds and with a minimum expenditure of energy. If you or I tried to do this we should confuse the horse and fail to get the 'tram' back. But watch a haulier and his horse. He has a special word of command. The horse knows that he has to give one sharp pull while his man puts his back to the 'tram' and guides.

The pit horse is intelligent and understanding. He knows his man. His man naturally cares for him. Even if the miner were not notoriously fond of animals; even if he were an inhuman creature he would be complicating his work by neglecting his horse or failing to establish contact with him.

Coal haulage is not one man's job; it is man plus horse. You have only to see — as you will always see in a mine — the haulier with his arm round his horse's neck, talking to him and patting him, to know that the Welsh pit horse is not a mere beast of burden.

'And,' said the farrier, 'in the morning most hauliers take something down for their horses. It may be only a lump of sugar; but there is generally something in their pockets — even in these hard times!'

I went 'down the line' with a new idea of the miner and with the knowledge that I can never again look at a bucket of coal without remembering him, black as pitch and wet with sweat, bracing his body against the coal wall half a mile from daylight....

'Well, and what do you think of it?'

A set of white teeth and two white eyeballs gleamed at me.

'You are always in the firing-line....'

'We get used to it. It's got to be done. I wish a few more people would come and see us work. Cheerio!...'

The cage moved and shot up towards the world. It was like a resurrection. I did not mind how it banged or rattled. I knew that above was sunlight and green grass! Water began to drip into the cage. The darkness of the shaft lightened. The cage leapt up and stopped.

I had to close my eyes. Daylight was blinding. It was as if my eyes had been turned inwards. But when I opened them I saw that it was raining. But how good it looked; how clean and marvellous!

And when I walked out of the cage a friend laughed at my black hands and my black face; and I knew again why the men coming up have no time for the men going down.

D.J. Williams
Handling Scott

Dai Richards from Carmarthen was another unforgettable character, a great hulk of a man, about thirty years old, some part of his ample body always likely to be showing through his ragged clothes. Yet if you saw him rolling down the street towards you, you'd think he owned it and that it was a privilege to be acknowledged by him.

He was a fisherman by trade, spending his summer months beside the green pastures of the Tywi between his home town and the sea at Llansteffan. But it was only the urgent need for food, or a pair of trousers to replace the pair that was threatening to leave him, which would induce him, some dark evening, to take his coracle on his back and lower it gently onto the water hoping to feel the sudden pull of a salmon. And if he got one large enough to shake his frail vessel with the slap of its tail, he was unlikely to go further, a meal being all he was after. In the winter, though, the fishing season over, he'd find work in one of the nearby mining villages, a place of sheltered hearths and great fires, — and he dabbled at his work there as tenderly as he tickled the trout in the Tywi. I first came across him during a night shift in the Lower Betws Colliery where he was handling Scott, a sturdy little grey mare whose notions of work were very like her master's. For if Scott took it into her head that a particular job was not worth tackling, she'd lower her nose to the ground, turn her ears back and stiffen her shoulders, standing firm as a rock for her principles, neither stick, whip nor hellfire curses managing to move her. Dai, of course, understood her only too well.

And it was in this happy mood of mutual understanding that I first encountered them, — Scott with an empty dram behind her, still as Lot's wife, on a less-than-steep incline, and Dai with perfect acceptance of the situation, seated comfortably on a pile of stones, his lamp between his feet, awaiting the renewed stirrings of his partner's spirit.

'Right then, little Scott,' he said patiently after some time, 'if you think you can get by like this, I'm damn sure I can. Take your time, my little maid.' And there in the bowels of Betws mountain I left the two in a profound study of the great problem of time.

The dialect of the Lancashire miners whom the company employed at this time was difficult to understand, and Dai, seeking to justify Scott's stubbornness and consequently the small amount of work the two accomplished, would say, 'You see now, they bring the Lancies and all these other buggers here as handlers and finally the old horses don't understand either English or Welsh.'

Robert Morgan
Accidents to Ponies

There were also accidents to ponies. An accident which was connected with a certain pony has stayed with me in all its detail. In fact it was not an accident at all, but a deliberate and planned episode which resulted in the death of a pony. This may sound callous and in one sense it is. On the other hand it can also be said to be a rough kind of justice carried out in the heat of emotion after a fatal accident to a fellow workman.

The episode of the pony begins with the absence of a certain haulier. The name of the pony was Sam, a frisky animal at times, but quite controllable when driven by an experienced haulier. On this particular shift Sam's haulier was absent and the pony was given to a collier to drive. This often happened, for some colliers, who volunteered to drive a pony, were quite experienced in handling such animals underground. The collier whose name was Davies (his Christian name I have forgotten) took upon himself the task of driving Sam. This he did successfully until midway through the shift. Davies stopped the pony in a narrow section of the road in order to insert sprags in the front wheels of the tram. The sprags acted as brakes, thus preventing the full tram from running out of control. As Davies placed one of the sprags in the wheel Sam pulled forward, trapping Davies under the tram from which the unfortunate man received fatal injuries. This fatality left the men angry and dazed, especially as it was the second time Sam had been directly involved in an accident to his driver.

When the news arrived of the accident to Davies, my father crawled out of the coal face and instructed me to gather up all the tools and place them on the bar. This I did as quickly as possible and locked the bar. I was pleased the shift had come to an end so early, but at the same time my thoughts went out to Davies whom I believed to be injured and not killed. The truth came to me a little later when we were back on the road and dressing to go out. One of the hauliers came into our road and spoke to my father. The news he brought was stunning and as I knew Davies and met him almost every day in the pit and often in the village streets, such news made the reality of the tragedy even more disheartening.

I was shocked and the coal dust on my face must have hidden the colour of shock.

We strolled to the mouth of the road and were joined by the same haulier who advised us to wait as they were preparing to take Davies out of the pit. A few minutes later another haulier passed by with Sam. He was leading the pony towards the face of the heading. I was confused as to what was happening. The accident had occurred on our heading and when I looked down the heading there was not a light to be seen. The carriers, I thought, were already on their way out with Davies.

'What's happened?' I eventually asked my father. Before I received an answer a voice called down from the heading road-head, 'keep clear ... down ... there!' The haulier who was with us stepped further back into the road. My father also stepped back a few paces and motioned me to do likewise. There was a rush of cold air and the rumbling of a tram. My father and the haulier listened and waited as the rumbling grew louder. Suddenly a pony and tram flashed past our road and went on and on down into the darkness of the heading. The tram was piled high with rocks and was running out of control, with no sprags in the wheels, and driving the pony Sam before it.

'That was Sam, wasn't it?' I asked and without waiting for an answer I scrambled out on the heading and stared into the darkness after the pony and tram. The rumbling sound died away and there was a sudden crash followed by silence.

'You've killed him!' I shouted, suddenly realising what had happened.

'He killed that chap Davies didn't he?' said the haulier, 'and injured that other collier not so long ago.'

Later when we strolled down the heading we came to a stout post which had been placed across the heading and against it lay Sam with the tram of rock spilled over him. Several workmen were clearing the rock and Sam lay quite still, half standing half lying. The tram was distorted and bent and it was obvious it had done its work on Sam. By now I felt sick with fear and sadness and my hands trembled, although I fought against such feelings and gave no evidence I had them. As we walked along we gradually caught up with the group of men who were carrying Davies on a stretcher. It was a slow walk and I helped to carry the lamps of those who gave a hand with the stretcher carrying. Later that day, when I had bathed and eaten a meal, I discussed the episode with other boys. I discussed it over and over again, being overwhelmed by the death of a man and a pony in one day.

Tony Curtis
Throwing the Punch

If you want to know about a man, watch how he treats animals. Or promote him over his mates. Give a man power and he'll use it or abuse it according to his nature. In all my years underground I learned that. Put a man in the dark and put the squeeze on him then watch out for the bad blood to come up in a well.

Take the fight between McIntosh and Whistler on the night shift in Lady Margaret, oh, thirty years back or more it was, but as clear as day to me now. Mining was a protected industry, the war effort, Dig for Victory, coal against the Nazis. So we was working double hard. They drafted hundreds of lads into the mines at that time. Whistler was one of those Bevin boys. Only he wasn't called Whistler from the start. His name was Richards I think, though it's so long since he was called that at work I couldn't be sure. Big enough for a lad — a pretty good shape on him really, so you couldn't have guessed there'd be trouble. But trouble there was — as soon as his turn come up at the top. As the cage clanked back up and the first men came out after their shift, Richards — Whistler — started to shake. Duw, the sweat came out of him — just like someone had turned on a tap. It was running down his face and neck soaking his shirt. You could hear his teeth, but he didn't say anything. I was next to him and gave Will Peters the other side a nod. We pressed him on both sides and managed to shuffle him into the cage. When the gate locked he seemed to let go and loosen against us. That's when the screaming started. All the way down we was deafened. They could hear us coming at the bottom like the bomb that's got your name. He fell out of the cage and Will Peters, having the room to swing now, laid such a bloody slap across his face. And that did it. Of course there was no question of him going up to the face. A man like that would be too much of a liability at the coal-face. That's where the money's made, and those boys worked their stalls like little empires. No, the foreman got word and Whistler was put on the horses. That's how he got his name. The boy had a way with them, talking them calm (talking himself calm, I suppose, as well) and whistling the whole time.

We must have had three hundred horses, maybe more, underground at that time in the Lady Margaret. Mining's always been a cruel industry. Back in the real old days they used to use children for hauling, and women, sometimes pregnant even, it didn't matter to the owners. They turned a blind eye to everything except the tally sheets. Keep the drams coming out steady and full, was all that mattered.

Anyway, each of these horses had a name — Albion, Betty, Crusader, Dreadnought, right down the alphabet and back again. And they all had housing in stables underground. Once down there a horse would never see the light of day again. Better, more humane in a way, I suppose, because once they got used to the gloom of the pit their eyes would never have taken the full glare of the sun again. It's bad enough for a man coming up out of that hole after a shift and taking the strength of the daylight.

Whistler didn't have any horses in his background; I found out talking that he was just another Dai from Pontypridd whose old man had a corner shop. But he seemed happy with them. 'They're my real wages,' he'd say, 'I'm happy talking to them..' He was so happy that after a week or so he'd almost bounce down in the cage. Mind you, he never went up to the face. Stuck to the horses and looking after them. He'd still be there today if it hadn't been for McIntosh.

Now, McIntosh was the son of a Scotsman who'd come into Wales back in the twenties to scab during the lock-out times. You know, everything changed after 1926. It were a good sense of comradeship you had in the mines until then. Afterwards the whole business went sour in the mouth. Debts and rumours and hatred and back-biting. You could see it in the eyes narrowing, like everyone was looking for the next move from someone.

Well, McIntosh followed his father into the pit and was always a big man for the money. I know what they say about the Scotch people, and a lot of rubbish it is to tar a whole country with the same brush I'm sure, but this McIntosh could have carried the whole blame on his shoulders. A real mean bastard he was, and doing anyone down that he could. he used to work times at the slaughter-house in the next village, a big bugger with plenty of muscle in his arms, but a tyre of beer belly pushing out against his buckle-belt too.

If Whistler was considerate and caring for the pit ponies, then he was alone in that way of being attached to them. Men who'd never dream of kicking a dog in their back-lanes would kick out at a horse in anger underground. Coal-mining is a hard frustrating business and with the pressure on you constantly to fill your drams and make your rate for wages, tempers can be spilled easier than coal. You daren't let fly at another collier though. The two things that are absolutely taboo down below are matches and punches. One or the other gets you your marching orders, sharpish.

I'm not saying as this McIntosh was the only one, but he was renowned for his cruelty. Not in short blasts of temper, but slow, lingering, unnatural. Like I'd seen him take a rat and hold it with his boot against a rail so a truck would cut it living in half. Well, this one night shift there was a sort of tension in the Lady Margaret. Two men had gone down under a fall during the day and one of them was near to death with a broken neck it was said. The Penderyn District had come across bad workings and more gas than usual too. Sometimes when you cut into the seam it breathes and groans back at you. Some will tell you it can sing, or play like organ pipes — weird it is. Anyway, the feeling was that the sooner the end of the shift and the end of the week came the better.

McIntosh was handling a horse called Tudor, a tired lump of bones that poor beast was too, four tins of dog meat and a pint of glue as they say. Likely as not the horse was on the back-end of a double shift, for although the ponies were brought back to stables for cleaning down, feeding and watering after a shift, quite often it was that the tally wasn't held to and within minutes the poor animal was being drawn out again by a miner on the next shift. A man pushing to get his wages up to a living level isn't going to look too closely at the only horse he's offered. Sometimes a beast would work a double-header then, the second shift miner demanding as much from the animal as his first handler.

Tudor was pretty well done for, but there's nowhere to run to for a horse down there in the dark. He was craning his neck down to the feed-bag that was hung low down around him. That meant he was distracted from his job of pulling and McIntosh wasn't making the time with his drams that he wanted.

'Come on ya lazy bugger! Move ya carcase!' McIntosh punching the flank of the beast and lashing out with his boot. And the horse leaning into the harness and struggling to take the incline with a loaded dram. Well, it seems McIntosh called a halt and was bending down to wedge the wheels firm on the slope when the horse eased the weight and the back wheel came close to depriving the miner of his fingers.

'Jesus, I'll 'ave you this time!' and instead of laying more kicks on the horse he goes cool-dangerous and takes loose the feed-bag from Tudor's neck. Uncoupling the iron shackles from an empty truck he adds them to the bag and replaces it around the horse's neck. Now those metal shackles must weight a mighty bit; just think of that lumped onto an already tired horse. It brought the beast's head down to breaking point and the full dram to be taken up the incline too.

God know how he made it, but he did, and the bulk of the rest of the shift as well. Then, when McIntosh has settled down for his food-break, tucking into his snap-tin like a pig, he leaves the weight on the horse and keeps him from the water too.

Now whether it was that Whistler happened that way or whether someone had dropped word of what McIntosh was up to, I don't know, but anyway, along he comes as if nothing was out of order.

'How's me Tudor then? Weary old boy, eh?' patting the horse along his flank and no doubt feeling the welts as he went. Then he goes to fasten up the feed-bag.

'Leave it be!' snarls McIntosh.

'Entitled to a feed like you or me,' reasons Whistler, and then,

'What's this, though? Where's his proper meal? There's a ton in this bag. What's your game, man?'

'What's it to you, boy? Mind ya own!'

'Help me get this thing off, you cruel bastard!' said Whistler, struggling with the bag.

'Leave it bloody well on and see to your own business!' and McIntosh got up from his snap. ''E's my horse this shift and he'll work to my ways, the good-for-nothing knackered lump!'

Whistler had managed to loosen the strap on Tudor's bag and was removing the weight when the Scotsman's hand grabbed his shoulder. Whistler must have put everything into swinging round and planting his fist in McIntosh's face for the bigger man went down like an axed steer, blood spread from his nose and mouth and he coughed on it. Whistler's hand wasn't so good either after that blow and he completed the job of watering the horse mostly with his left hand. Welcome that water was for the beast too, but he hardly moved apart from his drinking head. It must have been all so much of a muchness to him after those years down the pit — the damp, the explosions, the curses, the constant low roofs paring down his back.

There was McIntosh in the dust, wiping his sleeve across his face and spitting out, 'Bastard horse! Bastard horse!' over and over, low and bitter. Whistler had turned away to the horse again, and was unshackling the animal from the truck.

'There fella, there Tudor, boy.'

But at that moment along comes the overman. The thing about overmen is their noses. They can smell trouble from the other end of the pit.

'Something up here, then?' he says. 'Bit of trouble?'

'Nothing, boss,' says the Scotsman, struggling back to his feet and trying to turn his head away from the light.

'Anything up for you, Whistler?'

'Old Tudor, here, be on his last legs. Looks to me like he's done a double shift. No point killing the horse for a bit more coal. He needs changing over.'

'Oh, I see. Well, that's all right then, isn't it?' says the overman, turning to go. And then, 'Caught a piece of rock have you, Jock?'

'Stupid bloody pit prop,' came the mumbled reply. It was hardly uttered with conviction, but Roberts the overman had more sense than to push it. Give a man a chance to spit out his grievance, and then let it be was the order of the day.

Word got round that shift and Jock gave his notice in that week. He left for the Jubilee pit over in the next valley but didn't stick it long. He moved over to slaughterman full-time. He had a real forte in that area, no doubt.

Whistler was a quiet, good worker and I wouldn't mind betting he did some good with his influence over caring for the horses. Certainly, he had no call to bloody anyone's nose again that I heard of. But mining is no sort of vocation if you've got any sense and the price of a bus ticket, and Whistler was away as soon after V.E. Day as he could. I'd like to think of his going on with the horses, at a stable maybe, a brewery dray or settling into an old trade like blacksmithing. But that's just sentimental. He's probably on the buses in some city, or maybe worked himself up to clerking in an office and counting the days to retirement.

Sometimes we go all our lives pulling back from the thing we ought to do. But whatever happened in the long run, Whistler had his moment. You can't take that away — throwing the punch.

Gwyn Thomas
The Pot of Gold at Fear's End

Most of the outcrops driven into the mountainside by the men in their search for free coal were level workings, driven straight in. Not so the working of Naboth Kinsey. Naboth's enterprise was in a narrow cleft of the hillside and not much noticed. This suited Naboth for he was an obstinate, secretive man, not given to the long idle arguments about method that were loved by most of the outcrop workers, especially my father, who was a noted theorist, so noted he had already been thrown out of my brother Dan's outcrop for being a nuisance and holding up the work. So it was with great interest we heard that my father had been taken on as

an auxiliary at the working of Naboth Kinsey. Everyone said my
father and Naboth would get on very well, as well as a busy flame
in a deep petroleum well.

Naboth had scorned the level working adopted by most of his
comrades. He believed in going down as perpendicularly as he
could to contact the seams worked in the pits themselves. When
it was suggested to Naboth that this might take a long time, he
would take off his cap, play with his fingers on the top of his bald
head and say he did not feel rushed. He had lived for long enough
in the neighbourhood of mountains to know that it saves a lot of
burns on the inside of the skull to dim the light of immediate
thoughts by dipping them into some shadowed apprehension of
the everlasting. If you were going to do a job said Naboth, be
thorough. So he went on with his perpendicular digging. Some
said he would reach coal about fifty years after the coming of the
New Jerusalem and by then he would have got so far down he
would not be able to hear what the voters were saying who were
telling him that there was no longer any need for so much digging.
Others said he would probably strike a layer of salt and end up
talking about vinegar with Lot's wife. But the majority simply held
that one day Naboth and his companions would go down and find
it too much trouble to haul themselves back up again. So, when
it was announced that my father was teaming up with Naboth
there were quite as many people to feel sorry for my father as felt
sorry for Naboth. In the latter camp, my brother Dan was a kind
of president.

During his first few days with Naboth my father was full of
enthusiasm.

'You're like me, boy,' he told Naboth. 'You've got a feeling for
where the seams are. Me too. To us it's simple. The coal is
somewhere underneath us, so down we go, straight as a plummet.
That's fine deep hole you've got there, Naboth. It's straight, boy,
and it's deep. You didn't waste your plummet there. Perhaps it's
a little too deep from one point of view but there we are. The
seams are down there and we are the boys who go straight at them.
No messing. When do you think you'll hit the main seam, Nabe?'

'There's no hurry, Eli. We'll take it nice and steady,'

'That's the right spirit. That's the talk I like to hear. There are
chaps working on this mountain who'd spit their hearts up with
grief if they couldn't have a certain amount of coal in their hands
every day. The romance of this antic doesn't seem to appeal to
them at all. There's a lot of beauty that they don't see in just
making a hole. God knows where we'll get to before we finish

going straight down like this. They say the inside of the earth is just like the pan in a chip shop. You know those pans, Naboth, bubbling all the time. It would be a good idea if we got to that heat before we get to the seam, to take some potatoes down and do our own chips and cook them on the lava or whatever it is that does this bubbling.'

'We'll take it steady,' said Naboth. 'There's no hurry.' Not even the thought of chips which was the favourite food of the valley could stir this Naboth from his calm. It struck us that this man was a digger strictly for the sake of digging. Ends seemed to matter to him not at all and he would probably have started digging this hole in the bedroom floor if he did not have a wife and a bed that needed holding up.

To us, Naboth's little pit was every bit as dangerous as the level working of my brother Dan and if my father, with his quick eye for danger of all sorts, had not noticed this, we were sure it was only because he was still keen to show Dan what a way he had with him in this matter of tracking down seams; either that, or the swift, passionate flight of his own desires found something to detain and fascinate them in the serene, objectless methods of Naboth. Naboth's companions, Windsor Ellis and Elias Thompson, were like Naboth, quiet, sad-looking men who went about their work without any zest or relish as if glad that at least they were in no doubt that one day soon the whole issue would cave in around the whole pack of them ridding them, without fuss or expense, of air and trouble. The gear they had rigged up to get the diggers in and out of the hole looked most insecure. A bucket on a rope was let down from a cross bar. The digger got into the bucket and his two colleagues, three counting my father, took the strain on the rope. The earth was loaded into the bucket, hauled up by one man and emptied by the third. Windsor, whose moods, cradled in as mossy a nest of life-long mishaps as the valley could show, often took a bitter turn, regarded my father as frivolous and treated him with caution, especially if my father was anywhere near the rope and it was his, Windsor's, turn to go down into the diggings. But with Elias Thompson my father had much better luck. Elias was a man who had spent his whole life dominated by a woman of narrow religious tendency who had converted their bedroom into a centre meet only for prayer and bleak decency, numb with deep solitude and fitting texts. From this chaste cranny Elias peeped out at my father and found him to be, by his standards, a king of the goats and he looked upon his most harmless remark as a sin-soaked novelty to be stored away in some mental cupboard where not even his wife's probing life-hatred could pry. Later, it would be taken out to warm the frozen fragment of some thoughtful night.

My father's close friend Waldo Treharne, was depressed by the venture. Every time he looked at Naboth's pit-head gear with its rope and bucket we could see his mind painting a frame of doom around them as the shadow fell across his thoughts. It was clear that in terms of calamity he viewed Naboth's whole outfit as a fitting pendant to the bodily trouble, a hernia, which kept him idle.

'It was a bad step teaming up with Naboth,' he told my father. 'After one week of this, Eli, you will be praying to be back with the ponies.' My father in normal times was a stableman in the pit and of no great skill at this trade being stamped on and kicked as often as the floor by the little horses with which he had been trying for twenty years to come to some sort of understanding. 'It wouldn't surprise me,' went on Waldo, 'to find that Naboth is in league with Richards the Undertaker. I bet Richards calls on Naboth nightly to ask him when he can expect the big coop. Come to think of it, I saw Richards last night lurking by the door of the Library and Institute keeping an eye on you. Taking a rough measurement, no doubt. It would be more honest by a mile if Naboth threw aside all such dishonest tomfoolery as that bucket and tied the rope direct around your necks.'

'Waldo,' said my father very gently, 'your view is darkened by that trouble of yours, the hernia. It's pulled all your hope and joy clean out of shape. Naboth is a man to watch.'

'I'll do that, Eli. He won't be around to watch for very long. Nor you.'

Naboth kept my father out of the hole for several days and put him on the rope with Elias. This labour half killed my father who had done little in the past more strenuous than warding off the ponies and picking himself up when he couldn't. But he began to ache furiously from the strain of pulling the rope tight when Naboth or Windsor went down and his face went red as a sunset when it came time to haul them back up again. After his first day on the rope his body seemed to become fixed in the posture of rope-tugging and he walked as if he were being carried on a chair, his legs bent forward and his spine bent backward almost parallel with the ground. We followed him and Waldo home and it was very interesting to see my rather performing this Chinese bend and Waldo leaning as far forward as my father leaned back. They had to keep adjusting their step to be able to talk to each other at all.

Later that same evening, my father did what he always did when he found that life had once again put its foot upon his neck. He went to the Workmen's Library in search of a book that would provide him with an answer to this animal labour he was called upon to do at Naboth's hole. He found it. It was called *Through Breathing, Strength*. He brought it home excitedly, still followed by Waldo with a look of even deeper wonder in his eye. He stayed up reading it far into the night. Our bedroom was directly about the kitchen where he read. Off and on, we could hear a sound like that of a man drawing his shoe sharply across rough matting. That was my father filling up with breath. Then there would be ten minutes of coughing, choking, stumbling and swearing. That would be my father trying to rid himself quickly of breath that had got into parts of his body where breath had no right to be and where breath had never been before.

But the next morning he seemed cheerful and confident and we heard him tell Waldo on the way up to the outcrops that he had taken in enough of this breath doctrine to make a trial trip and that, in confidence, he now thought that most of Naboth's troubles were over.

'There are two ways of taking that, Eli,' said Waldo.

At the first opportunity my father took Elias aside and explained to him the advantages of this new system of taking the load off the muscles and putting it on the lungs. It took my father a long time and a few rough drawings to show Elias that this meant something more than simply tying the rope tightly around his chest. But Elias had faith in my father. He took it all in but owned up that he had no head for book work, had gone through life in a one-style sort of way, like a horse, and would stand little chance of making his own labour less by means of the new system.

'Once a horse, always a horse,' said Elias forlornly, and he said it so often, my father, the most bruised and aimed-at stable-man of his age and weight, stood clear, as if expecting Elias to fall in with the other ponies and throw a hoof at him.

'When I'm on that rope, Eli, pulling,' said Elias, 'I find it difficult enough breathing in the ordinary way without trying anything different.'

'You'll get into it, Elias. Patience and practice!'

'All right then. But don't say anything about this to Windsor or he'll stop us breathing altogether. When he's in the bucket he likes a steady hand on the rope. And he doesn't think so much of you as it is.'

'Windsor's a savage. He'll die toiling.'

It was when they were hauling Windsor up that morning that my father made his first experiment. We watched him closely. His eyes were closed. He scarcely seemed to be breathing at all and he wore around his lips a soft but masterful smile which he was copying from the man, a dark, wise-looking voter, whose photograph had appeared in the front of the book on breathing. The grasp of his hand on the rope was light and it seemed to pass through his fingers smoothly and without effort. At first we thought my father had got on the right track and the system was working. Then we heard a terrible gasping from Elias. His face was purple with overstrain and he was almost collapsing under the burden of pulling the whole load himself. When Windsor was safely delivered and out of earshot my father bent over the collapsed Elias. 'It works, Elias. It's a marvel, boy. Your mind seems to settle down into quiet sleep and you seem strong as a lion though you don't seem to be making any effort. And look at you. You stick to the old ways. You stick to muscle and brute strength. And look at you. You can hardly talk. You're a greater wreck than my friend Waldo who is also wedded to ancient ways. Now when Windsor goes back again, you try my way. Just control your breathing and put your mind on something that makes your mind feel good, anything but the rope.'

Windsor re-entered the bucket. He smiled, a brief, rare smile, at Waldo who was staring at him with eyes that were full of the most startling interest and pity. To Waldo, who had heard my father's words to the helpless and enchanted Elias, Windsor was now little more than a fly winging directly to its doom in the dark web of my father's dreams. As soon as Windsor's head sank out of sight, my father turned to Elias and said, 'Now, Elias. The breath. Hold it, boy.' Elias did that. He did it deliberately. His glottis jerked up like an arm, as if his breath were a fleeing dog that had to be caught before it could be held.

'Watch them now,' said Waldo tensely. A pattern of agony began to pull at the browskin of Elias. His mind was obviously seeking that something that would make it feel good and in all the gutted wilderness within his skull there was nothing that did not shrink away from his questing thought, weeping its still tininess midway between great good, great evil. Then his face began to change colour and he pointed at his mouth to tell us that he no longer seemed able to breathe at all. He looked terrified. He started to writhe and let go the rope. My father was taken by surprise. He himself, as senior breather and mentor in the team, had been leaving it all to Elias. He was jerked forward at tremendous speed by the descending mass of Windsor who was now hurtling out of control through the last few feet that separated him from the bottom of the shaft. The bottom was soft mud. It did no harm if you fell clear of the bucket. My father had some of the skin knocked off his chin when it smacked with shocking squareness against the crossbar of the hauling gear and Windsor, on reascending swore himself hoarse. Beyond that, no harm was done.

That afternoon, Windsor, in general discussion with Naboth, insisted that the one place where my father could not put the whole enterprise in danger was at the very bottom of the shaft, doing some digging. My father protested that this was foolishly wasting a high talent, now that he was well on his way to becoming a specialist on the rope. He hinted that if Windsor would stop being a clod and feel an urge to experiment they would soon see perfected a method of human haulage which would give Naboth's winding gear, now no more nor less than an organized pain in the the bowels to all concerned, the force and efficiency of the steam-driven units that lined the bed of the valley. Windsor, in no mood for dickering, stood over my father, snarled and said he would have none of this. And into the bucket my father had to go. As he cocked his leg to enter he wore a look which from one side recalled Columbus and from the other an old, worn gnome who knows that he is definitely on his way to his last frolic. Even as he descended, he kept on orating until the mumble of his voice vanished into the thick sucking clay of the walls.

We gathered around the top of the hole to see how he was getting along. We had been invited to do this by Naboth who was wanting to know why it was that after so many minutes no bucketful of earth had yet been sent aloft by my father and Windsor had shouted down to him to know what in hell's name he thought he was supposed to be doing down there. We peered down. After we had got him sorted out from the clay we could see him clearly enough. He was sitting at the side of the excavation, doing nothing but looking petrified and staring up, apparently at the sky.

'He's gone jingles,' said Windsor, putting his hand to his temple and beginning to wind, to leave us in no doubt as to where he thought my father had gone.

'He's frightened,' said Waldo, 'Eli has funny nerves.'

'Good God,' said Elias. 'Is that what it is? Look at his eyes. They are filling the hole. He must be holding his breath again. Tell him to breathe, Waldo. He looks horrible.'

Then my father started to shout. He shouted with a piercing suddenness that sent Elias bouncing back from the shaft like a

ball. 'It's coming in,' roared my father. 'Waldo! Elias! It's coming
in. Get me up! I'm trapped! Get me up!'
   They told him to get into the bucket. He fell silent instantly and
hopped in. They hauled him to the surface. He spent the rest of
his day explaining what had happened. As soon as he got down
there, he had looked up. And what did he see: Nothing. He said
it again. Nothing. We looked as if we did not believe him. Not a
damned thing, he repeated. Blank. Nothing. The sides of the hole
had seemed to slope inward until they met and he thought he was
trapped and shortly to be entombed and choked, the sides crumbling in upon him, with Windsor, still vindictive, shovelling a bit
of extra stuff down from above to make sure. There was a name
for the panic he had felt, he said. He had seen it in a book which
had described this thing driving people mad by the thousand. But
no book could really tell you what the hell of a feeling it was. It
was written on his heart and unless we wanted to wriggle down
his throat and have a look there the horror of it would have to
remain unread. 'This is another worry I've got to face,' he said,
after he had rested a while. 'Now I know I'm not to be trusted in
any hole where I'm supposed to be able to see the opening and
can't.' He grew bitter about this. 'Afraid. We're all afraid of
something always. Life is black and lousy with fear. Night is only
the stuff that rises from all the fear we sweat out of us through the
day. It shouldn't be. I'm going back down there. I've been afraid
of too many things. Bailiffs, bosses, dreams, neighbours, now
holes. But I'll conquer this fear it it's the last thing I do.'
   'It will be,' said Waldo, sombrely.
   The next day Naboth decided to give my father one last try. He
prepared to descend. Naboth, Elias and Windsor manned the
rope. Before entering the bucket my father rested his arm on the
cross-bar of the winding gear. Windsor began to mutter. My father
put on that soft but masterful smile again that he had picked up
from the dark-faced breath-controller in the book, as if to say to
Windsor, victim of toil, knew not what he did. He told us that he
could not let this moment pass without a few words of explanation
from himself. He then gave us a long survey of fear through the
ages and the manner in which it had nibbled upon the fibre of the
poorer voters like a rat upon cheese. He described himself as a bit
of cheese so nibbled he would have to bribe the average rat with
a bonus if he wanted to shed a few more crumbs. He told us of
the deadly effects fear had had upon the lives of his listeners, blaming
upon it such diverse complaints as Waldo's hernia, Naboth's bald-
ness, and Elias' wife. They all became very interested and by

grunts, shrugs and nods agreed that this fear was a lowering thing to be having about the place, a thing ripe to be shown the culvert. It was clear that my father was treating himself to a course of intense auto-propaganda to get his courage to the peak and when he arrived at the passage where he passed his own fears under review and he marked himself down as being descended from a long line of shudders, his self-pity welled up to a rising rhythm. It welled up so far he failed to see that Naboth, Elias and Windsor, thinking that this address would see them safe for at least another fifteen minutes, had dropped the rope and were following his argument with a dour mournfulness, as if it were a hearse. My father, thinking to make his exit dramatically, with the spotlight of his audience's sympathy still upon him, stepped quickly into the bucket and hurtled to the bottom at about a hundred miles an hour.

Windsor clambered down the rope after him. We all manned the rope to bring him, with my father on his shoulder, to the surface. My father had fallen nimbly and after a minute's hard slapping by Windsor, he opened his eyes. Naboth, who was a first-aid man said there was nothing broken and went pulling at my father's every limb as if he were disappointed about this and wanted to put it right. But nothing, not even Naboth trying to get everything out of the socket, could persuade my father to rise. He had had the wits scared out of him.

'I'm finished,' he kept saying. 'My doom was coal. But there's nowhere I'd have wanted it to happen better than here on my native hills.' We could see him sucking the pleasure from those words.

Then we called Dan and on to Dan's barrow we loaded him. We wheeled him home, watching the look of total, stricken bemusement on his face. Waldo explained to Dan how my father had taken up with the breathing caper.

'He learned it from an Indian in a book.'

'Old fakir!' bawled Dan as we came into the main street. 'Old fakir, one and four a sack. Cheap, cheap, buy now, buy now!'

My father waved the barrow to a halt. He stared piteously at Dan, got out and walked, offended to the very root of his strange, wondering self.

W.W. Gibson
The Last Shift

The gate clangs and the nightshift cage descends;
And, with eyes closed against the dust and grit
That swirls up in the draught, into the pit
Once more he drops, he, with his boyhood's friends,
Old mates and cronies now this many a year,
Packed close about him; and thinking, too, maybe,
Of their sons serving in the war, as he,
Of his own lad. For, as they drop down sheer,
Down, down and down, a thousand feet or more,
Down, down and down and down into the black
And tortuous entrails of the earth, young Jack,
A pilot since the outbreak of the war,
Happen, even now, is climbing three miles high
Or thrice three miles, up, up into the rare
And icy upper reaches of the air,
Up, up and up into the brilliant night
To tackle enemy squadrons, bearing down
To pound with death some sleepy English town—
Jack, soaring through thin air in flashing flight,
As into the thick closeness of the earth
His father drops, to work nightlong and hew
The coal, Jack, fighting...
              Yet maybe it's true
His own work, too, is fighting; for a dearth
Of fuel for the machines, without a doubt,
Would lose the war for us. Ay, sure enough,
Even planes could never soar unless the stuff,
Metal, and coal to smelt it, were dug out
Of earth's black bowels by such men as he,
The miner-sons of miners, who know the trick
Of handling tools, cutter and wedge and pick,
Almost by instinct.
              And now suddenly
At the shaft-foot the cage stops with a jerk
Beside the lamproom, and he takes his lamp,
Burnished and newly-tested against blackdamp;
Then mounts a tub to rattle to his work
Over the jolting trolley-rails and ride
Six miles or so along a gallery,
Long stript of coal, to where, beneath the sea,

Still richly-loaded measures run — the tide
Sweeping and surging in a welter of white
Far overhead, the island-circling deep
Where restless trawlers and destroyers keep
Unwinking watch throughout the livelong night....
And over them, the sky where, full of pluck,
Jack fights!
          Nay, he must not let his mind run
On suchlike thoughts! Jack is their only son;
But Jack, as other men, must take his luck.
And even in the pit.... Where should he be,
Himself, if he let his thoughts loose, sniffing all
The risks, the hundred things that might befall?
Life, at the best, was chancy: though, certainly,
War has increased the hazards: and even his wife,
Lying now snug in bed, God knows what might
Drop down on her from out of the clear night!
But he could not let his thoughts.... And such was life
For all of us in these days; everywhere
Folk faced such hazards, knowing that each breath
Might be their last: ay, all hobnobbed with death,
Hail-fellow-well-met! by sea or land or air.

'Twas strange tonight, though, how his thoughts had
On dangers. Ay, and reaching the pithead,
He had felt like turning back again, instead
Of stepping into the cage as he had done
So often without giving it a thought,
As if he fancied he might break his neck!
And, taking his lamp and handing in his check
To the lampman, old Dick Dodd, he had even caught
Himself out, muttering 'So long!' to him,
As though he would not see his old mug again,
Or cared much if he didn't! It was plain,
Plain as Dick's mug — and that was something grim —
His wits....
          His wife slept snug — Jack, overhead,
A red-haired guardian angel on the alert!
And, likely enough, neither would come to hurt
Tonight: and in the morning from her bed
His wife would rise as usual. For no wars
Could keep down Susan, always game and gay
To get things done. Even the Judgment Day

Would likely find her singing at her chores.
Ay, she would rise as usual to prepare
His breakfast and his tub and set things straight,
Against his coming. She was never late;
And he would always find things fair and square
On his return from the pit.
        And, as for Jack —
His folk had been pitfolk time out of mind;
And it took something special to down that kind
Or get them windy, even when things looked black.
Hazard was in their blood. They lived on risk,
And relished it, or took it as it came.

And now he hears somebody shout his name
Above the racket of the tubs; and brisk
And sharp he turns to answer an old jest —
He, always more than a match for anyone
When it came to ragging — while the trucks still run
Through the low dripping dusk, to come to rest,
Reaching their journey's end, with squealing brakes.
Then, nimbler yet than any, down he leaps;
And, scrambling over rocks and coaldust heaps,
And splashing through black puddles, now he takes
His way yet further along the narrow seam;
Stooping yet lower as the roof slopes down,
Rock-studded, threatening to crack his crown,
For all his leather cap; and wades a stream
That trickles from a rift in the coal-face.
Then, nigh on hands and knees, 'twixt closing walls
Into a three-foot seam he slowly crawls
And by his own coalcutter takes his place.

Crouched all night long, he works with aching bones,
Half-blind with dust and sweat: while all around
He hears the pit 'talk' as the stresses shift
And cutters grinding with harsh rasping sound,
While now and then a rattle of falling stones
Strikes sharply in his ear. Throughout the night
His thoughts are with his folk — his wife, asleep,
He trusts, in well-earned slumber, snug and deep;
And Jack above the clouds in reckless flight.
All night he works till, as the shift at last
Draws to an end, the cutter jams; and now,

Stopping to wipe a trickle from his brow,
He hears a long low rumble down the drift
That thunders nearer and nearer.... Roofs and walls
Heave all about him, cracking.... Blast on blast
Shatters his world for him ... till gradually
A dreadful quiet settles; and, by falls
Of rock cut off from life, he finds himself,
Together with his old mates, Bill and Joe,
Half-stifled, blind and dazed, as they crouch low,
Huddled in darkness on a narrow shelf.
Speechless, they crouch through an eternity;
Then, chuckling brokenly, he mutters, 'Come, Bill,
Let's clear our throats and turn a tune, until
They find us — and you, Joe! What shall it be?
Come, lads, pipe up! And, happen, they may hear,
And reach us easier.' Huskily, 'The Keel Row'
He starts; then, shyly joined by Bill and Joe,
His voice through the hot dark rings true and clear.

The Women

Anon.
The collier laddie

I've been east, and I've been west,
    And I've been in St. Johnstone,
But the bonniest laddie I ever saw
    Was a collier laddie dancing.

I've been east, and I've been west,
    And I've been in Kirkcaldy,
But the bonniest lass that ever I saw
    She was following a collier laddie.

She had silken slippers on her feet,
    Her body neat and handsome.
She had sky-blue ribbons on her hair
    With the gold about them glancing.

'Where are you going, my bonnie lass,
    And what is it they call ye?'
'Bonnie Jeannie Gordon is my name,
    And I'm following my collier laddie.'

'I'll give you gold, and I'll give you gear,
    And I'll make you my lady.
I'll make you one of high degree
    'Stead of following a collier laddie.'

I'll not have your gold, nor yet your gear,
    And I'll not be your lady.
But I'll make my bed in a collier's home
    And lie down with my collier laddie.'

When seven long years had come and gone,
    Seven long years or hardly,
This same gentleman came begging his bread
    From her and her collier laddie.

'Oh where is your gold, and where's your gear
    That were going to make me a lady?
Now I've got silver and gold enough
    From following my collier laddie.'

Idris Davies
Put Your Arms Around My Body

Put your arms around my body
And feed upon my breast,
And let your sorrow fade away
Into the darkening west.

In this hollow of the moorland
Young man, lie down with me,
And lose the day and its squalor
In the swoon of ecstasy.

Forget for tonight the tumult
The malice and the fret,
And know of the balm of my body
And clasp me and forget.

The hills and the vales are silent
And silent the stars above,
And my bosom is warm and gentle,
Young man, lie down and love.

Gwyn Jones
Mary

Mary had never loathed Jenkinstown as she did at that moment. She
knew now that she must get her feet out of the slime at once. She
saw the village composed entirely of smutty old gossips who hated
everything young and bright — dirty old men, dirty old women.

'Any idea, Mary?'

It would have to come out now. 'Yes. Sometimes when Mr.
Broddam goes to London on the Company's business, he has to
take his secretary with him.'

'I see,' said Oliver. 'Natural enough, too. But you have not been?'

'No. But I've got to go in a fortnight.'

'Oh,' said Oliver. 'You got to go in a fortnight.'

'Yes.'

'You didn't tell us, Mary.'

'No, mam. I would have told you this week-end, though.'

There was a short silence during which no one felt easy, and

then Oliver spoke as after deep thought. 'You think you are wise
to go, Mary?'

She looked at him with half-feigned amazement. 'Wise?'

'Well ay — wise.'

'Why ever not, dad?'

'Well, I donno quite — it seems to me that p'raps it might be
better not to.'

'Better, dad?' She laughed antagonistically. 'What choice shall
I have? Can a secretary say she doesn't want to go?'

'He can't make you, can he?'

'He can sack me, I suppose!'

Then Oliver made a decisive blunder. 'Well, you won't want to
stay there so very much longer, if he do.'

'Why, dad,' she cried, as startled as if her dismissal rested with
him; 'you don't think I want to finish, do you?'

'Well, you'll be getting married one of these days — or so I've
been thinking, and everybody else for that matter. You can't keep
on your job then, can you?'

'No?' she exclaimed incautiously. 'I must please myself about
all that!'

'And Edgar, Mary!'

'No, mam! I'll please myself and no one else.'

Insubordination in the home was a new experience for Oliver.
He had little liking for it. 'Don't talk so foolish,' he interrupted
sternly. 'You'll make a fine collier's wife with a typewriter down
in Newport, won't you?'

'Perhaps, then, I'll not be a collier's wife,' retorted Mary, now
on her feet by her mother.

'What's that!'

'I don't see why I shouldn't live my own life how I want to,'
cried Mary, near tears. 'All this old fuss!'

'Not be a collier's wife!' repeated Oliver. 'What d'you mean?'

'What I said!'

'I see,' said Oliver bitterly. 'You think yourself too good, I suppose?'

'Don't argufy, the pair of you!' Polly reproached them. 'It do
only mean a lot of old upset.' She held Mary's hand, and rubbed
her fingers.

'Upset enough! Ain't you goin' to marry Edgar?'

'Oh, don't bother the girl, dad! Let it rest till to-morrow.'

'I won't then. Not marry a collier, indeed!' Oliver felt that his
own flesh and blood despised him. 'Who d'you expect to marry
— this here Broddam?'

'You say that!' Mary panted. 'You dare to say a thing like that!

You dare! You are like all the rest of the Jenkinstown people, are
you? And you shan't speak to me like that! I won't allow you to
speak to me like that!'
    'That's enough! I'll speak to you how I like, my girl.'
    'You won't then!'
    Polly pulled herself upright. "Ave u no idea between u but to
argue, argue, argue? Go upstairs, Mary — go on, now.' Mary
yielded to the plea, and went from the kitchen with a defiant,
tearful look at Oliver. 'Oliver — you are old enough to know
better! I'm surprised at you — really I am.'
    'It ain't that, mam. I've 'eard too much to-night, I have, not to
wonder.'
    'Then you are wasting your time, Oliver. I 'ont hear a word agen
my Mary, so there.'
    She went out after Mary, leaving her husband to listen to her
lumbering tread on the stairs, and to brood miserably and angrily
over his own thoughts. But Polly went to Mary's room, and there
found her daughter face downwards on the bed, crying her eyes out.
Wisely, she let her cry for a time without attempting to comfort her,
and soon she quietened until she shook with only an occasional sob.
    'Well, Mary, there have been a proper old rumpus to-night.'
    'Oh, mam,' cried her daughter, snatching at her hand, 'I hate
everybody in Jenkinstown. I wish I was dead, I do!'
    'There now. You don't hate your mam, do you?'
    'No, mam — not you! You know I don't!'
    'I know that all right, Mary.' Sitting on the bedside, she held
her daughter. 'We all been upset, and that's the truth.'
    'That gossip, mam!'
    'When it do come, look, someone have got to suffer, Mary.
That's how the world do go.'
    'And dad believes it, I know. He does, doesn't he?'
    'No, he don't a bit. I know ewer father by this time.'
    Soon they were talking quite calmly. Polly, with more sense than
Oliver, knew that Mary was certain to go to London, and made
no attempt to argue her out of it. 'You know it's all right, mam,
don't you?' Mary asked at last.
    'Of course I do. I never give a thought to it. Go you, Mary. But
this about Edgar — did you mean that?'
    'I believe I did.'
    'You believe? You ought to know, Mary.'
    'Mam — will you listen if I tell you? I knew you would really,
and I've been wanting to tell someone ever since I started thinking
about it, only I was afraid somehow. Mam — I can never live in

Jenkinstown again.' The elderly woman and the young one sat there for half an hour in the dark, while Mary cleansed her bosom of much perilous stuff and explained to Polly all her fears and hopes. Strangely, Polly, who until this evening had never given such matters a thought, felt glad and relieved that her daughter had determined never to marry a miner. It was paradoxical that she could look back on a happy married life, and yet one that had never been free from worry and toil and pain. It was like destiny. Every month you saw it: a young girl marrying, strong and happy, then breaking, breaking, breaking; all the cares of the kitchen, the family, the pay-ticket; the never-ending round of washing, scrubbing, cooking, clearing away, polishing; the constant inflow of dirt; the child-bearing in agony after conception without desire and gestation without longing; brats at the breast, brats at the heels, brats at the apron-strings, a damning procession of life-drainers. At the best a life of denial and poverty, at the worst degradation. And always the indifference or contempt of your betters. Animals, to Mrs Shelton and her like; so much raw material to the coalowners; so many breeders of slaves. And always left to rot in drab and shabby villages. And those who go furthest down fester in squalidity, not blameless. But if your husband drank, was a sport, had a bad place in the pit, lost his health or his job, it was easy to come to this, easy to become pus in the running sore of humanity. Thinking confusedly of this, Polly uttered her usual thank-offering: 'Thank u, God.' It would never overtake her now; Mary was safe; pray God Luke would do well by his Olive.

The two women went downstairs together. 'You can't very well spend your time in your bedroom.' Polly told her sensibly, and Oliver was content to leave ill alone. When Luke came in, the household seemed quiet enough, and for a time he did not guess that their home life could never be the same again

William Hornsby
The Coal Owner and the Pitman's Wife

A dialogue Aa'll tell ye as true as your life,
Between a coalowner and a poor pitman's wife.
As she was a-walkin' alang the highway,
Whey, she met a coalowner and this she did say.
Chorus. Derry down, down, down derry down.

'Good mornin', Lord Firedamp,' this woman she says,
'Aa'll dee ye nee harm, sir, so divvent be afraid.
If ye'd been where Aa've been the most of mi life,
Whey, ye wadn't torn pale at a poor pitman's wife.'
Chorus.

'Then where de ye come from?' the owner he cried.
'Aa come from hell,' the poor woman replied.
'If ye come from hell, then come tell me reet plain,
How did ye contrive for te get oot again?'
Chorus.

'Whey, the way Aa come oot, sir, the truth Aa will tell,
They're tornin' the poor folk aall oot o' hell,
And it's this te mak' room for the rich, wicked race,
For there is a greet number of them in that place.'
Chorus.

'And the coalowners themselves is the next on command.
For to arrive in hell, so Aa understand,
For Aa heard the auld divil say as Aa come oot,
That the coalowners all had received their rout.'
Chorus.

'And how does the auld divil behave in that place?
Oh, sir, he is cruel to the rich, wicked race.
He is far more cruel than you can suppose,
Even like a mad bull with a ring through his nose.'
Chorus.

'If ye be a coalowner, sir, tek' mi advice,
Agree wi' your men and give them a fair price.
For if'n you don't, whey, ye knaa very well,
That ye'll be in greet danger of gannin' te hell.'
Chorus.

Good mornin, good woman, I must bid ye farewell,
For ye give me a dismal account aboot hell,
And if this be all true that ye say unto me,
Aa'll get hyem like a whippet, wi' me good men agree.
Chorus.

'Whey, the pit gates is closed, little mair Aa've te say,
Aa was torned oot mi' hoose on the thirteenth of May,
But it's noo te conclude and Aa'll finish mi sang,
Aa hope ye'll relieve me and let us carry on,'
Chorus.

Glyn Jones
Dock

Big sidings by the swing-bridge have black tons
Of trucked rock packed in twenty rows, sun-slavered
Coal converged upon the setting sun.
A black rat swims across the brass canal.

The sky tilts suddenly, its sleety herringbone
Of pouring rain spills thick across the dock,
Shags up the furry liner's side, above the dock-wall,
Pelts the sheety concrete, sprawls its gusty growths,
Its hiss of cold grey grass, across the tingling streets.

The gold coal-owner's daughter saves her hennaed
Hooks of hair, ducks pippin-breasted, slim grey-
Squirrelled, spindle-heeled, beneath some
Broker's porch; two big coal-trimmers, night-shift
Candle-pounds tucked dry, get after her.
One grins, 'God lummy Charlie', and is silent.

Only one watcher herring-gull, turning
High above the wetted town, sticks out,
His windy shoulder skilful in a storm
He saw ruled blue behind the Islands hours back.

Joe Corrie
Miners' Wives

We have borne good sons to broken men,
    Nurtured them on our hungry breast,
And given them to our masters when
    Their day of life was at its best.

We have dried their clammy clothes by the fire,
    Solaced them, cheered them, tended them well,
We have watched the wheels raising them from the mire,
    Watched the wheels lowering them to Hell.

We have prayed for them in a Godless way
    (We never could fathom the ways of God),
We have sung with them on their wedding day,
    Knowing the journey and the road.

We have stood through the naked night to watch
    The silent wheels that raised the dead;
We have gone before to raise the latch,
    And lay the pillow beneath their head.

We have done all this for our masters' sake,
    Did it in rags and did not mind;
What more do they want? what more can they take?
    Unless our eyes, and leave us blind.

Richard Llewellyn
Eating in the Strike

I held out bread and cheese, crusty bread, and yellow farm cheese, with cress and lettuce from the garden.

She looked at it, and swallowed again, with her hands behind her.

'Come on, girl,' I said. 'There is slow you are.'

So she took them and bit into them, and bit and bit and bit, until her little mouth was sure to burst, and her eyes had tears, and as she chewed she sobbed. Your throat goes dry and you cannot have your food in peace when somebody is hungry and shows it. So Shani had dinner for two of us that day, and in the afternoon she fell to the floor during history, and Mr Jonas carried her outside, and she went home with another girl.

I told my mother when I got home but she said nothing, only clicked her tongue and looked tired. There were many in the village just the same. Next morning I went to school with my can packed tight, and more in a brown paper parcel hanging on my coat button. Not a word from my mother, only a little smile.

So Shani and me sat together to have dinner every day in my mother's smile. I never saw her mother or father, and never went home with her, and though I asked her, she never came over in our Valley because they had sold their trap and it was too far for her. And after we had been having dinner for a couple of weeks she stopped coming. It was said that her father had gone to find work in the north, at Middlesbrough.

I will always remember her in something of blue and three lines of yellow braid, and a little bow on top of her hair and her face so pale and looking from the side like the face of the queens on coins of Ancient Greece.

Mr Gruffydd went time and time again to the Town and then back with food, money, and clothes for the people down in the hovels. But the people of the Hill would never have any of it. He was thinner, and his clothes were loose on him, and my mother said he would have starved if people had not asked him to eat with them, for he was paid by the Chapel, and the money was all gone in food for the hungry ones.

From all the men in idleness he got together a choir, and made Ivor second conductor to him. All over the valleys they walked singing for funds, and presently men from other valleys were coming over the mountain in dozens and scores to join them.

One night I heard a choir of a thousand voices singing in the darkness, and I thought I heard the voice of God.

Then children began to die.

The processions over the mountain were long at first and sometimes two or three a day. Then they grew shorter, and the hymns fewer, for the people had no strength.

July, August, September, October.

November.

The cold was on us and snow was thick in the very first week. People were burning wood, and some of the men went down to the colliery to get coal and were stopped by the watchmen, but they took no notice and loaded up. Next morning, police came by brake, and went to live in the lamp house. Two men who were caught were taken over the mountain and had six months in jail. So those who had no money for coal went up on the mountain for wood, and since all the people in other valleys were looking for wood, there was soon no wood to be had, except standing trees. But they were green and not to be lit by anything but a fire.

More and more children were dying, and now women were dying, and men. No more were coffins built by Clydach. A sheet had to do, and did.

Two, three, and four families went into one house to eat and have warmth together. Windows were boarded to keep out cold.

Even Mr Gruffydd had trouble to keep the men from a riot, and going down to the Colliery and killing the police.

One morning in the third week, Ellis the Post stopped Mari outside our house and gave my father a letter.

'Come you in, Ellis,' said my mother. 'Breakfast is ready.'

'No, indeed, Mrs. Morgan, my little one,' Ellis said, and pale about the nose with cold. 'I will have it when I get home, see.'

'You shall have breakfast now,' said my mother, 'or never come inside this house again.'

'Yes, Mrs Morgan,' said Ellis, and off with his cap, and sitting next to me. 'But no tea, and no bacon, if you will excuse me.'

'Tea you shall have, and bacon, and potatoes,' said my mother, and ready to fly at him. 'And please to have what you are given.'

'Yes, Mrs Morgan,' said Ellis, and hang-dog, with his eyes looking at her upwards and sideways.

'If there shall come a time when you leave this house without a proper something to eat,' said my mother, 'look for me on the floor.'

'Beth,' my father said, and passing the letter to Ianto, 'the boys and me will go into Town to-day.'

My mother looked at him straight, with her fork in the potatoes and one foot on the fender.

'Well?' she said.

'The owners,' my father said, with more colour in his face than I had seen for weeks.

'We shall have to give it in,' Davy said, sipping hot water.

'They have promised a minimum wage,' my father said. 'That is a straw, at all events.'

'And we are the drowners,' said Ianto, looking at the letter still.

My father raised his fists and hit the table to make the crockery jump.

'No matter,' he shouted, with flames in his eyes. 'Let us drown, then. But by God Almighty, I will have food in those children's little bellies before the night is out.'

'Good, Gwilym,' my mother said. 'Go you. Angharad, go to Mr Gruffydd and ask him to breakfast.'

'O, Mama,' Angharad said, and jumped from the stool, and flew from the house.

We had a lovely breakfast that morning, indeed.

Bacon sliced thick, and potatoes, and toast with butter, and strawberry jelly, and tea, with sugar and milk, too. There is good it is to have good food with taste after a long time without.

'Where have you been hiding all this, Beth, my little one?' my father said, eating the breakfast of two and a pleasure to watch.

'You mind your affairs,' my mother said, and blushing red and beautiful, indeed, 'and I will please to attend mine. Have I been living all this time and nothing to show?'

'Beth, my sweet love,' my father said, 'you were made and the mould was hit with a hammer.'

'Go from here,' said Mama, tears and laugh together, 'before I will give you a good hit with one, too.'

Idris Davies
Mrs Evans Fach, You Want Butter Again

Mrs Evans fach, you want butter again.
How will you pay for it now, little woman
With your husband out on strike, and full
Of the fiery language? Ay, I know him,
His head is full of fire and brimstone
And a lot of palaver about communism,
And me, little Dan the Grocer
Depending so much on private enterprise.
What, depending on the miners and their
Money too? O yes, in a way, Mrs Evans,
Yes, in a way I do, mind you.
Come tomorrow, little woman, and I'll tell you then
What I have decided overnight.
Go home now and tell that rash red husband of yours
That your grocer cannot afford to go on strike
Or what would happen to the butter from Carmarthen?
Good day for now, Mrs Evans fach.

W.H. Davies
The Collier's Wife

The Collier's Wife had four tall sons
    Brought from the pit's mouth dead,
    And crushed from foot to head;
When others brought her husband home,
Had five dead bodies in her room.

Had five dead bodies in her house —
  All in a row they lay —
  To bury in one day:
Such sorrow in the valley has
Made kindness grow like grass.

Oh, collier, collier, underground,
  In fear of fire and gas,
  What life more danger has?
Who fears more danger in this life?
There is but one — thy wife!

Anon.
I'll Have a Collier

I went out to get some water,
Get some water for me tea,
And I caught me foot and down I stumbled,
The collier lads come kissin' me.

My mother said I mustn't have a collier,
For it would surely break her heart,
I don't care what my mother tells me,
I'll have a collier for my sweetheart.

If you leave your collier sweetheart,
I'll buy you a guinea-gold ring,
You shall have a silver cradle,
For to rock your babies in.

Well, I don't want your silks and satins,
I don't want your guinea-gold ring,
I don't want your silver cradle,
For to rock me baby in.

Collier lads get gowd and silver,
Ferranti's lads get nowt but brass,
And who'd be married to a lad from Ferranti's,
When there are plenty of collier lads.

My mother said I could be a lady,
If from me collier lad I'd part
But I'd sooner walk on the bottom of the ocean,
Than I'd give up me collier sweetheart.

And I went out to get some water,
Get some water for me tea,
Caught me foot and down I stumbled,
The collier lads come kissin' me.

D.H. Lawrence
Washing the Man

The clock struck eight and she rose suddenly, dropping her sewing on her chair. She went to the stairfoot door, opened it, listening. Then she went out, locking the door behind her.

Something scuffled in the yard, and she started, though she knew it was only the rats with which the place was overrun. The night was very dark. In the great bay of railway lines, bulked with trucks, there was no place of light, only away back she could see a few yellow lamps at the pit-top, and the red smear of the burning pit-bank on the night. She hurried along the edge of the track, then, crossing the converging lines, came to the stile by the white gates, whence she emerged on the road. Then the fear which had led her shrank. People were walking up to New Brinsley; she saw the lights in the houses; twenty yards farther on were the broad windows of the Prince of Wales, very warm and bright, and the loud voices of men could be heard distinctly. What a fool she had been to imagine that anything had happened to him! He was merely drinking over there at the Prince of Wales. She faltered. She had never yet been to fetch him, and she never would go. So she continued her walk towards the long straggling line of houses, standing blank on the highway. She entered a passage between the dwellings.

'Mr Rigley? — Yes! Did you want him? No, he's not in at this minute.'

The raw-boned woman leaned forward from her dark scullery and peered at the other, upon whom fell a dim light through the blind of the kitchen window.

'Is it Mrs Bates?' she asked in a tone tinged with respect.

'Yes. I wondered if your Master was at home. Mine hasn't come yet.'

''Asn't 'e! Oh, Jack's been 'ome an' 'ad 'is dinner an' gone out.
'E's just gone for 'alf an hour afore bedtime. Did you call at the
Prince of Wales?'

'No —'

'No, you didn't like —! It's not very nice.' The other woman
was indulgent. There was an awkward pause. 'Jack never said
nothink about — about your Mester,' she said.

'No! — I expect he's stuck in there!'

Elizabeth Bates said this bitterly, and with recklessness. She
knew that the woman across the yard was standing at her door
listening, but she did not care. As she turned:

'Stop a minute! I'll just go an' ask Jack if 'e knows anythink,'
said Mrs Rigley.

'Oh, no — I wouldn't like to put —!'

'Yes, I will, if you'll just step inside an' see as th' childer doesn't
come downstairs and set theirselves afire.'

Elizabeth Bates, murmuring a remonstrance, stepped inside.
The other woman apologized for the state of the room.

The kitchen needed apology. There were little frocks and trousers
and childish undergarments on the squab and on the floor, and a
litter of playthings everywhere. On the black American cloth of
the table were pieces of bread and cake, crusts, slops, and a teapot
with cold tea.

'Eh, ours is just as bad,' said Elizabeth Bates, looking at the
woman, not at the house. Mrs Rigley put a shawl over her head
and hurried out, saying: 'I shanna be a minute'.

The other sat, noting with faint disapproval the general untidi-
ness of the room. Then she fell to counting the shoes of various
sizes scattered over the floor. There were twelve. She sighed and
said to herself, 'No wonder!' glancing at the litter. There came the
scratching of two pairs of feet on the yard, and the Rigleys entered.
Elizabeth Bates rose. Rigley was a big man, with very large bones.
His head looked particularly bony. Across his temple was a blue
scar, caused by a wound got in the pit, a wound in which the coal
dust remained blue like tattooing.

''Asna 'e come whoam yit?' asked the man, without any form
of greeting, but with deference and sympathy. 'I couldna say
wheer he is — 'e's non ower theer!' — he jerked his head to signify
the Prince of Wales.

''E's 'appen gone up to th' Yew,' said Mrs Rigley.

There was another pause. Rigley had evidently something to get
off his mind: 'Ah left 'im finishin' a stint,' he began. 'Loose-all 'ad
bin gone about ten minutes when we com'n away, an' I shouted,

"Are ter comin', Walt?" an' 'e said, "Go on, Ah shanna be but
a'ef a minnit," so we com'n ter th' bottom, me an' Bowers, thinkin'
as 'e wor just behint, an' 'ud come up i' th' next bantle —'

He stood perplexed, as if answering a charge of deserting his
mate. Elizabeth Bates, now again certain of disaster, hastened to
reassure him:

'I expect 'e's gone up to th' Yew Tree, as you say. It's not the
first time. I've fretted myself into a fever before now. He'll come
home when they carry him.'

'Ay, isn't it too bad!' deplored the other woman.

'I'll just step up to Dick's an' see if 'e *is* theer,' offered the man,
afraid of appearing alarmed, afraid of taking liberties.

'Oh, I wouldn't think of bothering you that far,' said Elizabeth
Bates, with emphasis, but he knew she was glad of his offer.

As they stumbled up the entry, Elizabeth Bates heard Rigley's
wife run across the yard and open her neighbour's door. At this,
suddenly all the blood in her body seemed to switch away from
her heart.

'Mind!' warned Rigley. 'Ah've said many a time as Ah'd fill up
them ruts in this entry, sumb'dy'll be breaking their legs yit.'

She recovered herself and walked quickly along with the miner.

'I don't like leaving the children in bed, and nobody in the
house,' she said.

'No, you dunna!' he replied courteously. They were soon at the
gate of the cottage.

'Well, I shanna be many minnits. Dunna you be frettin' now,
'e'll be all right,' said the butty.

'Thank you very much, Mr Rigley,' she replied.

'You're welcome!' he stammered, moving away. 'I shanna be
many minnits.'

The house was quiet. Elizabeth Bates took off her hat and shawl,
and rolled back the rug. When she had finished, she sat down. It
was a few minutes past nine. She was startled by the rapid chuff
of the winding-engine at the pit, and the sharp whirr of the brakes
on the rope as it descended. Again she felt the painful sweep of
her blood, and she put her hand to her side, saying aloud, 'Good
gracious! — it's only the nine o'clock deputy going down,' rebuking
herself.

She sat still, listening. Half an hour of this, and she was wearied
out.

'What am I working myself up like this for?' she said pitiably to
herself. 'I s'll only be doing myself some damage.'

She took out her sewing again.

At a quarter to ten there were footsteps. One person! She watched for the door to open. It was an elderly woman in a black bonnet and a black woollen shawl — his mother. She was about sixty years old, pale, with blue eyes, and her face all wrinkled and lamentable. She shut the door and turned to her daughter-in-law peevishly.

'Eh, Lizzie, whatever shall we do, whatever shall we do!' she cried.

Elizabeth drew back a little, sharply.

'What is it, mother?' she said.

The elder woman seated herself on the sofa.

'I don't know, child, I can't tell you!' — she shook her head slowly. Elizabeth sat watching her, anxious and vexed.

'I don't know,' replied the grandmother, sighing very deeply. 'There's no end to my troubles, there isn't. The things I've gone through, I'm sure it's enough—!' She wept without wiping her eyes, the tears running.

'But, mother,' interrupted Elizabeth, 'what do you mean? What is it?'

The grandmother slowly wiped her eyes. The fountains of her tears were stopped by Elizabeth's directness. She wiped her eyes slowly.

'Poor child! Eh, you poor thing!' she moaned. 'I don't know what we're going to do, I don't — and you as you are — it's a thing, it is indeed!'

Elizabeth waited.

'Is he dead?' she asked, and at the words her heart swung violently, though she felt a slight flush of shame at the ultimate extravagance of the question. Her words sufficiently frightened the old lady, almost brought her to herself. 'Don't say so, Elizabeth! We'll hope it's not as bad as that; no, may the Lord spare us that, Elizabeth. Jack Rigley came just as I was sittin' down to a glass afore going to bed, an' 'e said, "'Appen you'll go down th' line, Mrs Bates. Walt's had an accident. 'Appen you'll go an' sit wi' 'er till we can get him home." I hadn't time to ask him a word afore he was gone. An' I put my bonnet on an' come straight down, Lizzie. I thought to myself, "Eh, that poor blessed child, if any body should come an' tell her of a sudden there's no knowin' what'll 'appen to 'er." You mustn't let it upset you, Lizzie — or you know what to expect. How long is it, six months — or is it five, Lizzie? Ay!' — the old woman shook her head — 'time slips on, it slips on! Ay!'

Elizabeth's thoughts were busy elsewhere. If he was killed — would she be able to manage on the little pension and what she could earn? — she counted up rapidly. If he was hurt — they wouldn't take him to the hospital — how tiresome he would be to nurse! — but perhaps she'd be able to get him away from the drink and his hateful ways. She would — while he was ill. The tears offered to come to her eyes at the picture. But what sentimental luxury was this she was beginning? She turned to consider the children. At any rate she was absolutely necessary for them. They were her business.

'Ay!' repeated the old woman, 'it seems but a week or two since he brought me his first wages. Ay — he was a good lad, Elizabeth, he was, in his way. I don't know why he got to be such a trouble, I don't. He was a happy lad at home, only full of spirits. But there's no mistake he's been a handful of trouble, he has! I hope the Lord'll spare him to mend his ways. I hope so, I hope so. You've had a sight o' trouble with him, Elizabeth, you have indeed. But he was a jolly enough lad wi' me, he was, I can assure you. I don't know how it is....'

The old woman continued to muse aloud, a monotonous irritating sound, while Elizabeth thought concentratedly, startled once, when she heard the winding-engine chuff quickly, and the brakes skirr with a shriek. Then she heard the engine more slowly, and the brakes made no sound. The old woman did not notice. Elizabeth waited in suspense. The mother-in-law talked, with lapses into silence.

'But he wasn't your son, Lizzie, an' it makes a difference. Whatever he was, I remember him when he was little, an' I learned to understand him and to make allowances. You've got to make allowances for them —'

It was half-past ten, and the old woman was saying: 'But it's trouble from beginning to end, you're never too old for trouble, never too old for that —' when the gate banged back, and there were heavy feet on the steps.

'I'll go, Lizzie, let me go,' cried the old woman, rising. But Elizabeth was at the door. It was a man in pit-clothes.

'They're bringin 'im, Missis,' he said. Elizabeth's heart halted a moment. Then it surged on again, almost suffocating her.

'Is he — is it bad?' she asked.

The man turned away, looking at the darkness:

'The doctor says 'e'd been dead hours. 'E saw 'im i' th' lamp-cabin.'

The old woman, who stood just behind Elizabeth, dropped into a chair, and folded her hands, crying: 'Oh, my boy, my boy!'

'Hush!' said Elizabeth, with a sharp twitch of a frown. 'Be still, mother, don't waken the children: I wouldn't have them down for anything!'

The old woman moaned softly, rocking herself. The man was drawing away. Elizabeth took a step forward.

'How was it?' she asked.

'Well, I couldn't say for sure,' the man replied, very ill at ease. ''E wor finishin' a stint an' th' butties 'ad gone, an' a lot o' stuff came down atop 'n 'im.'

'And crushed him?' cried the widow, with a shudder.

'No,' said the man, 'it fell at th' back of 'im. 'E wor under th' face, an' it niver touched 'im. It shut 'im in. It seems 'e wor smothered.'

Elizabeth shrank back. She heard the old woman behind her cry:

'What? — what did 'e say it was?'

The man replied, more loudly: ''E wor smothered!'

Then the old woman wailed aloud, and this relieved Elizabeth.

'Oh, mother,' she said, putting her hand on the old woman, 'don't waken th' children, don't waken th' children.'

She wept a little, unknowing, while the old mother rocked herself and moaned. Elizabeth remembered that they were bringing him home and she must be ready. 'They'll lay him in the parlour,' she said to herself, standing a moment pale and perplexed.

Then she lighted a candle and went into the tiny room. The air was cold and damp, but she could not make a fire, there was no fireplace. She set down the candle and looked round. The candle-light glittered on the lustre-glasses, on the two vases that held some of the pink chrysanthemums, and on the dark mahogany. There was a cold, deathly smell of chrysanthemums in the room. Elizabeth stood looking at the flowers. She turned away, and calculated whether there would be room to lay him on the floor, between the couch and the chiffonier. She pushed the chairs aside. There would be room to lay him down and to step round him. Then she fetched the old red table cloth, and another old cloth, spreading them down to save her bit of carpet. She shivered on leaving the parlour; so, from the dresser-drawer she took a clean shirt and put it at the fire to air. All the time her mother-in-law was rocking herself in the chair and moaning. 'You'll have to move from there, mother,' said Elizabeth. 'They'll be bringing him in. Come in the rocker.'

The old mother rose mechanically, and seated herself by the fire, continuing to lament. Elizabeth went into the pantry for another candle, and there, in the little penthouse under the naked tiles, she heard them coming. She stood still in the pantry doorway, listening. She heard them pass down the three steps, a jumble of shuffling footsteps and muttering voices. The old woman was silent. The men were in the yard.

Then Elizabeth heard Matthews, the manager of the pit, say: 'You go in first, Jim. Mind!'

The door came open, and the two women saw a collier backing into the room, holding one end of a stretcher, on which they could see the nailed pit-boots of the dead man. The two carriers halted, the man at the head stooping to the lintel of the door.

'Wheer will you have him?' asked the manager, a short, white-bearded man.

Elizabeth roused herself and came from the pantry carrying the unlighted candle.

'In the parlour,' she said.

'In there, Jim!' pointed the manager, and the carriers backed round into the tiny room. The coat with which they had covered the body fell off as they awkwardly turned through the two doorways, and the women saw their man, naked to the waist, lying stripped for work. The old woman began to moan in a low voice of horror.

'Lay th' stretcher at th' side,' snapped the manager, 'an' put 'im on th' cloths. Mind now, mind! Look you now!'

One of the men had knocked off a vase of chrysanthemums. He stared awkwardly, then they set down the stretcher. Elizabeth did not look at her husband. As soon as she could get in the room, she went and picked up the broken vase and the flowers.

'Wait a minute!' she said.

The three men waited in silence while she mopped up the water with a duster.

'Eh, what a job, what a job, to be sure!' the manager was saying, rubbing his brow with trouble and perplexity. 'Never knew such a thing in my life, never! He'd no business to ha' been left. I never knew such a thing in my life! Fell over him clean as a whistle, an' shut him in. Not four foot of space, there wasn't — yet it scarce bruised him.'

He looked down at the dead man, lying prone, half-naked, all grimed with coal-dust.

'"'Sphyxiated," the doctor said. It *is* the most terrible job I've ever known. Seems as if it was done o' purpose. Clean over him, an' shut 'im in, like a mousetrap' — he made a sharp, descending gesture with his hand.

The colliers standing by jerked aside their heads in hopeless comment.

The horror of the thing bristled upon them all.

Then they heard the girl's voice upstairs calling shrilly: 'Mother, mother — who is it? Mother, who is it?'

Elizabeth hurried to the foot of the stairs and opened the door:

'Go to sleep!' she commanded sharply. 'What are you shouting about? Go to sleep at once — there's nothing —'

Then she began to mount the stairs. They could hear her on the boards, and on the plaster floor of the little bedroom. They could hear her distinctly:

'What's the matter now? — what's the matter with you, silly thing?'— her voice was much agitated, with an unreal gentleness.

'I thought it was some men come,' said the plaintive voice of the child. 'Has he come?'

'Yes, they've brought him. There's nothing to make a fuss about. Go to sleep now, like a good child.'

They could hear her voice in the bedroom, they waited whilst she covered the children under the bedclothes.

'Is he drunk?' asked the girl, timidly, faintly.

'No! No — he's not! He — he's asleep.'

'Is he asleep downstairs?'

'Yes — and don't make a noise.'

There was silence for a moment, then the men heard the frightened child again:

'What's that noise?'

'It's nothing, I tell you, what are you bothering for?'

The noise was the grandmother moaning. She was oblivious of everything, sitting on her chair rocking and moaning. The manager put his hand on her arm and bade her 'Sh — sh!'

The old woman opened her eyes and looked at him. She was shocked by this interruption, and seemed to wonder.

'What time is it?' — the plaintive thin voice of the child, sinking back unhappily into sleep, asked this last question.

'Ten o clock,' answered the mother more softly. Then she must have bent down and kissed the children.

Matthews beckoned to the men to come away. They put on their caps and took up the stretcher. Stepping over the body, they tiptoed out of the house. None of them spoke till they were far from the wakeful children.

When Elizabeth came down she found his mother alone on the parlour floor, leaning over the dead man, the tears dropping on him.

'We must lay him out,' the wife said. She put on the kettle, then returning knelt at the feet, and began to unfasten the knotted leather laces. The room was clammy and dim with only one candle, so that she had to bend her face almost to the floor. At last she got off the heavy boots and put them away.

'You must help me now,' she whispered to the old woman. Together they stripped the man.

When they arose, saw him lying in the naive dignity of death, the women stood arrested in fear and respect. For a few moments they remained still, looking down, the old mother whimpering. Elizabeth felt countermanded. She saw him, how utterly inviolable he lay in himself. She had nothing to do with him. She could not accept it. Stooping, she lay her hand on him, in claim. He was still warm, for the mine was hot where he had died. His mother had his face between her hands, and was murmuring incoherently. The old tears fell in succession as drops from wet leaves; the mother was not weeping, merely her tears flowed. Elizabeth embraced the body of her husband, with cheek and lips. She seemed to be listening, inquiring, trying to get some connection. But she could not. She was driven away. He was impregnable.

She rose, went into the kitchen, where she poured warm water into a bowl, brought soap and flannel and a soft towel.

'I must wash him,' she said.

Then the old mother rose stiffly, and watched Elizabeth as she carefully washed his face, carefully brushing the big blond moustache from his mouth with the flannel. She was afraid with a bottomless fear, so she ministered to him. The old woman, jealous, said:

'Let me wipe him!' — and she kneeled on the other side drying slowly as Elizabeth washed, her big black bonnet sometimes brushing the dark head of her daughter-in-law. They worked thus in silence for a long time. They never forgot it was death, and the touch of the man's dead body gave them strange emotions, different in each of the women; a great dread possessed them both, the mother felt the lie was given to her womb, she was denied; the wife felt the utter isolation of the human soul, the child within her was a weight apart from her.

At last it was finished. He was a man of handsome body, and his face showed no traces of drink. He was blond, full-fleshed, with fine limbs. But he was dead.

'Bless him,' whispered his mother, looking always at his face, and speaking out of sheer terror. 'Dear lad — bless him!' She spoke in a faint, sibilant ecstasy of fear and mother love.

Elizabeth sank down again to the floor, and put her face against his neck, and trembled and shuddered. But she had to draw away again. He was dead, and her living flesh had no place against his.

A great dread and weariness held her: she was so unavailing. Her life was gone like this.

'White as milk he is, clear as a twelve-month baby bless him, the darling!' the old mother murmured to herself. 'Not a mark on him, clear and clean and white, beautiful as ever a child was made,' she murmured with pride. Elizabeth kept her face hidden.

'He went peaceful, Lizzie — peaceful as sleep. Isn't he beautiful, the lamb? Ay — he must ha' made his peace, Lizzie. 'Appen he made it all right, Lizzie, shut in there. He'd have time. He wouldn't look like this if he hadn't made his peace. The lamb, the dear lamb. Eh, but he had a hearty laugh. I loved to hear it. He had the heartiest laugh, Lizzie, as a lad —'

Elizabeth looked up. The man's mouth was fallen back, slightly open under the cover of the moustache. The eyes, half shut, did not show glazed in the obscurity. Life with its smoky burning gone from him, had left him apart and utterly alien to her. And she knew what a stranger he was to her. In her womb was ice of fear, because of this separate stranger with whom she had been living as one flesh. Was this what it all meant — utter, intact separateness, obscured by heat of living? In dread she turned her face away. The fact was too deadly. There had been nothing between them, and yet they had come together. For as she looked at the dead man, her mind, cold and detached, said clearly: 'Who am I? What have I been doing? I have been fighting a husband who did not exist. *He* existed all the time. What wrong have I done? What was that I have been living with? There lies the reality, this man.' And her soul died in her for fear: she knew she had never seen him, he had never seen her, they had met in the dark and had fought in the dark, not knowing whom they met nor whom they fought. And now she saw, and turned silent in seeing. For she had been wrong. She had said he was something he was not, she had felt familiar with him. Whereas he was apart all the while, living as she never lived, feeling as she never felt.

In fear and shame she looked at his naked body, that she had known falsely. And he was the father of her children. Her soul was torn from her body and stood apart. She looked at his naked body and was ashamed, as if she had denied it. After all, it was itself. It seemed awful to her. She looked at his face, and she turned her own face to the wall. For his look was other than hers, his way was not her way. She had denied him what he was — she saw it now. She had refused him as himself. And this had been her life, and his life. She was grateful to death, which restored the truth. And she knew she was not dead.

And all the while her heart was bursting with grief and pity for him. What had he suffered? What stretch of horror for this helpless man! She was rigid with agony. She had not been able to help him. He had been cruelly injured, this naked man, this other being, and she could make no reparation. There were the children — but the children belonged to life. This dead man had nothing to do with them. He and she were only channels through which life had flowed to issue in the children. She was a mother — but how awful she knew it now to have been a wife. And he, dead now, how awful he must have felt it to be a husband. She felt that in the next world he would be a stranger to her. If they met there, in the beyond, they would only be ashamed of what had been before. The children had come, for some mysterious reason, out of both of them. But the children did not unite them. Now he was dead, she knew how eternally he was apart from her, how eternally he had nothing more to do with her. She saw this episode of her life closed. They had denied each other in life. Now he had withdrawn. An anguish came over her. It was finished then: it had become hopeless between them long before he died. Yet he had been her husband. But how little!

'Have you got his shirt, 'Lizabeth?'

Elizabeth turned without answering, though she strove to weep and behave as her mother-in-law expected. But she could not, she was silenced. She went into the kitchen and returned with the garment.

'It is aired,' she said, grasping the cotton shirt here and there to try. She was almost ashamed to handle him; what right had she or any one to lay hands on him; but her touch was humble on his body. It was hard work to clothe him. He was so heavy and inert A terrible dread gripped her all the while: that he could be so heavy and utterly inert, unresponsive, apart. The horror of the distance between them was almost too much for her — it was so infinite a gap she must look across.

At last it was finished. They covered him with a sheet and left him lying, with his face bound. And she fastened the door of the little parlour, lest the children see what was lying there. Then, with peace sunk heavy on her heart, she went about making tidy the kitchen. She knew she submitted to life, which was her immediate master. But from death, her ultimate master, she winced with fear and shame.

Joe Corrie
The Miner Lover

Here in the guts of the earth,
   In my father's tomb,
In the forests of aeons past,
   In the gas and the gloom;
Naked and blind with sweat
   I strive and I strain,
Like a beast in the famine year,
   Or a blood Cain.

But, home, I will wash me clean,
   And over the hill,
To the glen of the fair primrose
   And the daffodil;
And there I will sing of my Love
   With a tenderness
That only a god can feel —
   Lord God, what a mess!

Barry Hines
A Gold Signet Ring

The first ring of the telephone brought Forbes's head up so violently, that Stacey and the Mines Inspector heard something crack in his neck. He had the handpiece to his ear before the second ring. The others lost their fatigue too when they heard what he said.

'All four?... What about their checks?... Wait a minute, Tom.'

He found a pencil on his desk and started to write on a pad. It was obvious from the comments he made as he listened and wrote, that he was being told the names and condition of the trapped rippers.

'All right, Tom, we'll check that straight away. Keep me informed, won't you?'

He replaced the telephone, then picked up the pad and looked at it. The other two leaned forward in their chairs, anxious to hear what he had been told.

'They've found them. Two of them are dead and two still alive. They're in a bad way by the sounds of it, both unconscious, and they're receiving emergency treatment before they bring them out of the pit.... The trouble is they can only identify two of them; Albert Rhodes, he's dead, and the lad, Alan Dobson, he's still alive, thank God.... It's the other two, they've lost their checks, so before I ask anybody to come and identify them, we need to know who they are first. We don't want to make things worse by giving somebody's wife the wrong news.'

Still holding the pad, he stood up and walked towards the door.

'You'll have to excuse me a minute until I've got this lot sorted out.'

After he had gone, the Union Official and the Mines Inspector did not speak. They were numb from lack of sleep and waiting; and now that the missing men had finally been found, they were unsure how to react: the news was both good and bad.

Forbes and Beatson walked down the corridor towards the reception room. Beatson was listening intently to what the Manager was telling him, and when they reached the room, Forbes gave him the pad and went back to his office. Beatson stood holding the door handle for several seconds, rehearsing what he was going to say before he opened the door and looked inside. Only Kath and Mrs King were left in the room now. Beatson beckoned to Ronnie's wife and they went out in the corridor and closed the door.

'Is there any news, Mr Beatson?'

'Well, yes. We've found them, but we're not sure of their identities yet. We're just finding out who's who.'

He walked her away from the reception room just in case Kath could hear what they were saying.

'Tell me, Mrs King, was your Ronnie wearing a pair of green football socks at all?'

She looked mystified, then angry. it seemed like a macabre joke in the circumstances.

'Green football socks? What's he want to be wearing football socks for? He's never played a game of football in his life as far as I know.'

'What about a ring? Was he wearing a gold signet ring?'

She did not answer, and looked at him warily to discover the motive for the question. But his expression was ambiguous and she was forced to concede.

'Yes, he does wear a ring, why?'

'A square one, with a criss-cross pattern on one side?'

'That's right. His mother bought it him for his twenty-first birthday.'

Beatson stopped walking and turned to face her square on.

'Then I've got some bad news for you, Mrs King. I think you'd better come into Mr Carter's office and sit down.'

She stared at him, refusing to accept what he meant, then her fist went to her mouth and she collapsed as cleanly as if she had been knocked out. But Beatson was ready for her: he caught her and with his arm around her, supported her along the corridor towards Carter's room.

Forbes stood with Carter at the window of his office and watched the last ambulance leave the pit yard. It was light outside, and across the fields, the clouds above the village were being lit from below, as the sun came up behind the houses. Two reporters called to each other as they unlocked their cars, and a television crew packed their equipment into a van. Across the yard, the winding wheels slowed down until their spokes were visible again, and a few seconds later, the Milton rescue team walked slowly across the yard towards the baths, silent and exhausted.

Carter turned away from the window. 'Well, I suppose we'd better go and tell Albert's wife. There should be somebody up now.'

Forbes watched the Salvation Army officer empty a large metal teapot on to the ground.

'She's not in the house on her own, is she?'

'No, she's a daughter looking after her.'

Carter walked across to the door, then paused and turned back.

'When we've been, I think I'll get off home and get a bit of sleep, if that's all right?'

'Yes, that's all right, Geoff. I'll see you later.'

Carter left the room and Forbes sat down at his desk. He was too exhausted to work. He just sat there, staring and dazed, and Sheila had to knock twice on the door before he told her to come in.

'There's some telegrams here, Mr Forbes. They've just arrived.'

She passed them across the desk, and while he read them, she collected the plates and cups and stacked them on the coffee table in the centre of the room.

Janet opened her eyes at the sound of the kitchen door being unlocked. She wondered why she was on the settee; remembered, and jumped up as her mother and Tony came through into the living-room.

'What's happened, mam? Where's my dad?'

She went dizzy from getting up too fast, and had to support herself on the arm of the settee.

'They've taken him to hospital.'
'Is he all right?'
Kath sat down at the table.
'He's alive, that's the main thing.'
Janet put her hands to her face and started to cry.
'He's got a broken leg, broken pelvis, and he's badly burned on his chest and arms. I didn't see him. He was unconscious when they brought him out....'
This made Janet cry even more, and her sobbing made Tony angry.
'What are you crying for? He'll be all right now.... Better than five I could name, anyroad.'
Janet looked over her hands at him.
'Why, have some been killed?'
'Five. There's five been killed.'
The stairs door opened and Mark came through the kitchen and pushed past Tony into the living-room. He was wearing his pyjamas, and he scratched his body through the material as he looked at his mother and Janet.
'What's our Janet crying for? Where's my dad?'
Kath beckoned him over to her, and she did not tell him until he was standing by her and she could hold him.
'He's all right, love. There's been an accident at the pit and he's been injured. They've had to take him to hospital, but don't worry about it, he'll be all right.'
Mark pulled away from her so that he could see her face properly.
'Will he be better in time to take me to the Test Match next week?'
There was silence the room, then Tony laughed. If he had let himself go, he would have become hysterical and wept shamelessly. Janet was furious, and looked as if she was going to cross the room and give Mark a good hiding.
'You selfish thing! Is that all you can think about when he's nearly been killed?'
Mark, bewildered by his sister's outburst, looked to his mother for reassurance.
'I didn't know, did I?'
Kath pressed her lips together to stop herself from crying, then went across to the window and opened the curtains, so that the others would not see her tears. Tony switched off the light.
'We'll take you, Mark. Me and your uncle Harry'll take you.'
Mark looked at the tickets wedged behind the clock on the mantelshelf.
'It'd be better if my dad could go though.'

Two streets away, Carter stopped his car outside Albert Rhodes's house and got out with Alf Meakin. They stood on the pavement for a few seconds rehearsing what they were going to say before Alf knocked on the door. There was a pause, then Alf cleared his throat and took off his cap when he heard the bolts being withdrawn at the other side.

Glyn Jones
Marwnad

The little oil lamp burns between us;
    It is made of blue glass;
Beside the bare table we sit waiting
    For the night to pass.

In the other room, through the open door,
    I can see him lying dead,
With his pit clothes folded where they dropped them
    Beside the bed.

His mother watches the fleece of flames
    That grows over the embers,
Weeping softly for this, and those other griefs
    She half remembers.

My cheeks are dry for you, my man,
    But you know what's for me —
Even now I am wondering when my pains
    Will come upon me.

Joe Corrie
Women are Waiting Tonight

Women are waiting tonight on the pit-bank,
Pale at the heart with dread,
Watching the dead-still wheels
That loom in the mirky sky,
The silent wheels of Fate,
Which is the system under which they slave.
They stand together in groups,

As sheep shelter in storm,
Silent, passive, dumb.
For in the caverns under their feet,
The coffin seams of coal
'Twixt the rock and the rock,
The gas has burst into flame,
And has scattered the hail of Death.
Cold the night is, and dark,
And the rain falls in a mist.
Their shawls and their rags are sodden,
And their thin, starved cheeks are blue,
But they will not go home to their fires,
Tho' the news has been broken to them
That a miracle is their only hope.
They will wait and watch till the dawn,
Till the wheels begin to revolve,
And the men whom they love so well,
The strong, kind loving men,
Are brought up in the canvas sheets,
To be identified by a watch,
Or a button,
Or, perhaps, only a wish.
And three days from now,
They will all be buried together,
In one big hole in the earth.
And the King will send his sympathy,
And the Member of Parliament will be there,
Who voted that the military be used
When last these miners came on strike
To win a living wage.
His shining black hat will glisten over a sorrowful face,
And his elegantly shod feet will go slowly behind the bier.
And the director of the company will be there,
Who has vowed many a time
That he would make the miner eat grass.
And the parson, who sits on the Parish Council,
Starving the children and saving the rates,
Will pray in a mournful voice,
And tear the very hearts of the bereaved.
He will emphasise in godly phrase,
The danger of the mine,
And the bravery and valour of the miner.
And the Press

That has spilled oceans of ink
Poisoning the public against the 'destroyers of industry',
Will tell the sad tale,
And the public will say, 'How sad',
But a week today all will be forgotten,
And the Members of Parliament,
The coalowner,
The parson,
The Press,
And the public,
Will keep storing up their venom and their hatred,
For the next big miner's strike.
Women are waiting tonight at the pit-bank,
But even God does not see
The hypocrisy and shame of it all.

Tony Curtis
Preparations

In the valley there is an order to these things:
Chapel suits and the morning shift called off.
She takes the bus to Pontypridd to buy black,
But the men alone proceed to the grave,
Neighbours, his butties, and the funeral regulars.
The women are left in the house; they bustle
Around the widow with a hushed, furious
Energy that keeps grief out of the hour.

She holds to the kitchen, concerned with sandwiches.
It is a ham-bone big as a man's arm and the meat
Folds over richly from her knife. A daughter sits
Watching butter swim in its dish before the fire.
The best china laid precisely across the new tablecloth:
They wait. They count the places over and over like a
    rosary.

Strife: Strikes, Unions and Politics

Harri Webb
That Summer

The first thing I remember is the General Strike,
My father in his shirtsleeves leaning on the front gate
Smoking his pipe in the sunshine,
Miss Davies the shop calling across to him,
Are you out, Mr Webb? I hear now
Her bright amused voice, see Catherine Street
Empty and clean, hear the nine days' silence
As the last ripple of a lost revolution
Ebbed into history and the long defeat
Began to mass its shadows. The ambulances
Were absent from the road beside the hospital,
Garn Goch Number Three, Great Mountain, Gilbertsons,
Elba, the names I learnt to read by, names
Of collieries and tinworks, names of battlefields
Where a class and a nation surrendered
The summer they killed Wales.

We spent the time on the sands, played all day.
We had the whole place to ourselves,
Or so it has always seemed, from the West Pier
To Vivian Stream. When you are five years old
There are things you understand more easily
Than ever afterwards, that the sea is huge
And goes on for ever from Swansea, the moon
And the hospital clock inhabit the same sky,
Neighbours. But there are other things, and these
You only understand later, much later.
Inland, in those ambulance villages, the other side
Of Town Hill, from stations further up the line
From Mumbles Road, already it was beginning,
The losers' trek, the haemorrhage of our future.
But for a child there is only the present.

Dad, I said, there'd be lovely if the strike
Was all the time, then you and me could come
Down the sands every day and play. He laughed.
It wouldn't do, son, he said, it wouldn't do,
There's got to be work, see, there's got to be work.
Chasing a ball I didn't stop to argue, forgot
I'd ever asked the question till later, long after
The summer my country died.

Meic Stephens
Eels

With the colliers out a week now
the Taff at Rhydfelen is deep and as green
as my father says it's ever been,
and boys on the bridge, like I was once, throw
lines into the fast stream below.

Many's the time I stood by there
as the river slid with its load of small coal
past our village, over the weir,
all silver and smooth as a mole.
I'm wondering what our barbed ambitions were —

the pike with sovereign scales,
perhaps, or even the most venerable salmon
that swam in the legendary pools of Wales?
Fact is, nothing was ever won
by our bent pin and penny bun

from such poor water; except
during the troubles there were always eels
those brazen scavengers kept
us in full summer's sport as up they leapt
from the coiled darkness towards our reels,

and we felt the spun blood's
principalities generating through our rods
as the monsters, hugely fanged,
writhed in their long agonies. I understand
why, at last, like stripling gods

we would decapitate the devils:
they came to the cwm when times were slack,
the factory's gate and the foundry's stack
beset by their seething evils.
At school we used to paint our rivers black,

blue was for the sky and sea.
I remember that on the bridge at Rhydfelen,
hoping the boys will not see
this landed man's ferocity
as he turns to smash those wicked heads in.

Jack Jones
The Communist Procession

On his way up to the Bon Marché he was held up by the long and strong procession of Communists and unemployed which was marching to the meeting-place in the open air which was not a stone's-throw from the Workmen's Hall, where, at the same hour, the Miners' Federation and Rhondda Labour Party May-day meeting was to be held.

The Communist procession was headed by a 'Revolutionaries' Jazz Band' which was something more than merely funny; it was, at times, inspiring; and its bandmaster was none other than our old comrade Mordecai Rees, by this time also the leader of the Red Rebels Dance Band which was drawing young Rhondda to dance to Mordecai's class-conscious numbers at the Ambulance Hall on Monday, Thursday and Saturday each week, Admission 6d (ladies and unemployed half-price). His last two numbers, the 'Rostov Rag' and the 'Trotsky Trot', were being whistled by the young people in all parts of the Rhondda, and there was about this time a strong rumour to the effect that Mordecai and his Red Rebels were likely to receive an invitation from the Comintern to tour Russia under the auspices of the All-Russian Central Council of Culture Disseminators.

Close up to the band, under the red banner on which was inscribed 'Workers of the World, Unite', marched the redoubtable Dai Hippo, with the two speakers for the demonstration one on either side. One of the speakers was far gone in years, but the flashing eyes were the eyes of a youth. With fifty-five years of trade-union and revolutionary activity behind him, he marched like a well-trained soldier, commanding the respect of all who lined the streets — irrespective of party — for, whatever might be said of him, he had been a fighter, and a great one, all his days. The other speaker, also a fighter, was about half the size and age of his colleague and was well worth listening to even though he were not worth following.

Close on the heels of the speakers came the local leaders, Comrades Ike Matson, Lily Hopkins, Councillor Rose Morris who had got back on to the Rhondda Council at the March elections, Trevor Short, Jerry Dando, and others too numerous to mention.

Stretched out in fours for about a quarter of a mile behind the leaders were the rank and file of Rhondda Communism, many carrying banners on which could be read 'Down With the Means

Test', and 'Down' with many other things besides. A platoon here
and there sang, and there were many women carrying babies who
marched at their husbands' sides. A few of those lining the street
tittered now and then as the procession marched by, but Dan saw
nothing to laugh at. What he saw were rows of careworn faces,
faces which had looked in many directions for help (and guidance
before becoming bitter and turning to Communism as the last
hope.

Anon.
The Miners' Catechism 1844

1. *Ques. What is your name?*
Ans. PETER POVERTY.
2. *Q. Who gave you that name?*
A. My godfathers and godmothers in my baptism, wherein I was
made a member of the Black Coal Pit, a child of Slavery, and an
inheritor of the sunless mine.
3. *Q. What did your godfathers and godmothers then for you?*
A. They did promise and vow three things in my name. First,
that I should renounce all opposition to my master's will. Secondly, that I should believe that every Word and Action of the
Viewers was said and done for my benefit. Thirdly, that I should
obey them in every thing, work for their benefit alone, and live in
poverty and want all the days of my life.
4. *Q. Dost thou not think that thou art bound to believe and to do
as they have promised for thee?*
A. Yes, verily, and by God's help so I will not. And I am very
thankful that such a spirit of resistance is within me, for I see
plainly that they would bind me to slavery to my life's end.
5. *Q. Rehearse the articles of thy belief.*
A. I believe that my master, who is a Coal Owner, sinks his
capital in a coal pit in expectation of making a princely fortune in
a short time, and that his Viewers are hard task-masters, and that
it is them who make such hard laws as these: Thou shalt not send
any foul coal, splint, or stone to bank among the good coals, or if
thou dost thou shalt pay threepence a quart for it; and if thou
sends a tub to bank not containing the quantity specified in the
bond, thou shalt not have any pay for the same. I believe that the
Overman is a tool in the hands of the Viewers, and that he will do
anything the Viewers want him, such as not measuring the rank

of the Putters, cheating the men out of their yard-work, and enforcing the fines contained in the bond.

6. *Q. What dost thou chiefly learn in the articles of thy belief?*

A. First, I learn to believe that the Coal Owner wants his work done as cheap as possible. Secondly, that the Viewer gets as much money as possible from the workmen so that he may accumulate as large a fortune as possible. Thirdly, that the Overman will do anything the Owner or Viewer tells him to do.

7. *Q. You said that your godfathers and godmothers did promise for you, that you should keep your master's commandments, tell me how many there be.*

A. Ten.

8. *Q. Which be they?*

A. The same that the Master spake on the day that he read the bond, saying, I am thy Master who hath bound thee to the Coal Pits, and for twelve months thou shalt remain in bondage to me.

I. Thou shalt have no other Master but me.

II. Thou shalt work for no other Master, neither shalt thou give thy services to any other man, for I thy Master am a jealous man, and I will visit thee with heavy fines and punishments if thou break any of my commandments.

III. Thou shalt say no manner of evil of me, but shalt say I am a good master although I act as a tyrant towards thee; for I will not hold thee guiltless if thou say any manner of evil of my name.

IV. Remember thou work six days in the week, and be very thankful that I allow thee the seventh day to recruit thy exhausted strength, for I, thy Master, want as much work out of thee as possible, and if it suits me to give thee only two, three, or more days in the week, be very thankful that I give thee any work at all, for I only look at my interest, not thine.

V. Honour me thy Master, honour my Viewers, my Overman, and my Agents, so that thy days may be long in my service.

VI. Thou shalt work thyself to death and commit self-murder.

VII. Thou shalt exhaust thyself with work to hinder thee from committing adultery.

VIII. Thou shalt not steal anything from thy Master altho' I give thee no money for working for me.

IX. Thou shalt not bear witness against me or any of my agents for any misdemeanour we may commit.

X. Thou shall not covet thy Master's house, thou shalt not covet thy Master's wife, nor his servants, nor his lands, nor his carriage, nor his horses nor any thing that is his.

9. *Q. What dost thou chiefly learn from these commandments?*
A. I learn two things, my duty towards my master, and my duty towards myself.
10. *Q. What is thy duty towards thy master?*
A. My duty towards my master is to work honestly for him and not to waste his substance, while I remain his servant.
11. *Q. What is thy duty towards thyself?*
A. My duty towards myself is to work a fair day's work for a fair day's wage, and to hinder my master from being a tyrant over me.
Catechist. My good man you know your duty well; continue to pursue the path of duty, look well to your own interest and to that of your masters and you will do well; train up your children to hate tyranny, love freedom, justice, and truth; and may the God of Heaven and Earth bless you to your life's end.

A PRAYER

Unto they care and protection O most unmerciful Master, I commit myself this day. Preserve me from all fines, cheatery, deductions, either by weight or measure, or from any thing contrary to the justice of my labour; by thy *graceless assistance* enable me to receive my pay without any subtraction, division, or reduction from its true and rightful amount; and so conduct my *interest*, that neither sloth, idleness, drunkenness, nor the *ill-will-ness* of Overman or Viewer may occasion the neglect of it, that I may be dutiful and obedient to all Viewers, Owners, Masters, or Tyrants; that I may love them with that due submission and affection to which their dissimulation and tyranny so justly entitle them. Preserve me from all misfortunes and accidents. Leave me not to the wolfish and devouring fiends of despotism and *crush-down-ism*; but so guide me by common sense, that I may live without any fear of them now, or dread of them hereafter. Bless all those who are linked in the bonds of *Union*. Bless all those who *have*, or may relieve our necessities, and all associations of good feeling and humanity. Purge all collieries from the pestilential fever of proud, arbitrary, and domineering oppression. By thy condescension permit me to receive the full fruit of my labour this day, and defend me from all dark-dealing treachery in the night, so that I may be enabled to procure substantial food for myself and family, thro' the merits of my own diligence, industry, and perseverance. Amen.

Idris Davies
There's a Concert in the Village

There's a concert in the village to buy us boots and bread,
There's a service in the chapel to make us meek and mild,
And in the valley town the draper's shop is shut.
The brown dogs snap at the stranger in silk,
And the winter ponies nose the buckets in the street.
The 'Miners' Arms' is quiet, the barman half afraid,
And the heroes of newspaper columns on explosion day
Are nearly tired of being proud.
But the widow on the hillside remembers a bitterer day,
The rap at the door and the corpse and the crowd,
And the parson's powerless words.
And her daughters are in London serving dinner to my lord,
And her single son, so quiet, broods on his luck in the queue.

Tommy Armstrong
The Miners of South Medomsley

If you're inclined to hear a song Aa'll sing a verse or two,
And when Aa's done ye're gannin te say that ivory word is true.
Noo, the miners of S. Medomsley they nivor will forget,
Fisick and his tyranny and hoo they had been tret.
For in the midst of danger, these hardy sons did toil,
For to earn their daily bread se far beneath the soil.
To make an honest livelihood each miner did contrive,
But ye shall hear hoo they were served in eighteen eighty five.

Chorus.
Oh, the miners of South Medomsley, they're gannin te mek
    some stew.
They're gannin te boil fat Postick and his dirty candy crew.
The maistore should have nowt but soup as lang as they're alive,
In memory of their dirty trick in eighteen eighty five.

Below the county average then the lads was ten per cent,
But Fisick, the unfeelin' cur, he couldn't rest content.
A ten per cent reduction from the men he did demand,
But such a strong request as this the pitmen couldn't stand.

Their notices was all seemed oot and when they had expired,
All the gear was brought te bank and the final shot was fired,
To hurt his honest working men this law-lived man did strive,
But he'll often rue for what he did in eighteen eighty five.
Chorus.

Fisick was determined still more tyranny te show,
For te get some candymen he wandered to and fro,
He made his way up te Consett and he saw Postick the Bum,
He knew he liked such dirty work and he wad surely come.
Fisick telt him what te dee and where to gan and when,
And at the time appointed Postick landed with his men,
Wi' polises and wi' candymen the place was all alive
All through the strike that Fisick caused in eighteen eighty five.
Chorus.

Then Commander Postick give the word and they started wi'
    their work,
And tho' they were done at five o'clock they dursn't stop till
    dark;
And when they had done all they could and finished for the day,
The bobbies guarded Postick and his dirty dogs away.
Fisick was a tyrant and the owners were the same,
For the turn-oot o' the strike they were the men to blame,
Neither them nor Postick need expect they'll ivor thrive
For what they did te Dipton men in eighteen eighty five.

Ronald Ferguson
There's Ganna Be Some Trouble

That afternoon our usual gang gathered to play football, and we had just selected sides and kicked off the ball to start the game when we saw men and women hurrying towards the close mouth. Willie Douglas, the biggest lad of our gang, picked up the ball. 'Come doon and see whits gaun on. Ah doot there's ganna be some trouble.' We rushed to the close, to find a crowd of neighbours, men and women, blocking the entrance. There was excited talk. 'Aye, they'll stop the blacklegs alright,' shouted one woman. 'Whaur aboots hae ah these men cam fae?' asked another. 'Bloody hell, here's the bloody police, merchin down the street.' 'Hell, it wull be murder. Polis noo! If they start there wull be a bloody riot. The colliers ill noo staun fur that.'

Cha and I tried to worm our way through the crowd, and finally got right up to the front, but our success did us no good. Mrs Davidson spotted us. 'Hai, Davie, catch they twa wee buggers!' Davie caught me by the right ear and Cha by the backside of his trousers, and we were pushed and shoved back through the crowd out of harm's way. There was some laughter from the men and women, but we were angry at our unceremonious treatment. The gang laughed at us when we were once more at the back of the crowd, but questioned us eagerly about what was going on. We told them that the road was packed with people, but that the police were coming. 'There'll be a battle noo,' said Rab, 'and we'll not be able to see it.'

'Aye we wull,' I cried. 'If we gang ower the dyke inti Walker's building and oot the close at the tap we kin see it awe.' This was a good idea, and we whooped with joy. We raced to the dyke that separated the two buildings and clambered over, heedless of the shouts of some women who had spotted us, and knew what we were doing. We cleared the obstacle and ran through the open space that served as a drying green, ducking under the clothes that hung on the lines, with more shouts from the angry women. 'Watch ma claes, ye wee devils!' We reached the mouth of the close, to find only a few people there, watching what was happening further down the road. Men were coming out of the Miners' Rows and marching down towards the Drygate, past the two-storey building and stopping at the Tally shop we knew as Angelo's. The crowd grew by the minute, jostling for a position where each man could see down the Drygate, the route the blacklegs must come.

We stood by the mouth of the close, not knowing what to expect, but sensing trouble. There was a shout: 'Here they come, the blackleg bastards!' There were screams of hatred, obscenities yelled and a surge forward as the scabs approached, with their police escort. 'Git the bastards,' and there was a rush forward. As the two sides clashed, punches were thrown on both sides, and the police escort was broken up. There were several fights taking place at the same time, and more men were rushing forward out of the Rows and closes. Suddenly a mass of blue uniforms appeared from what seemed nowhere as police reinforcements charged into the fighting mass, truncheons flying. There were bloody heads and faces, but this seemed to enrage the miners even more, and they tore into the attack with boots and fists. Gradually, as more police appeared, the battle, which had been swaying backwards and forwards, swayed in favour of the police and the blacklegs, who had suffered the brunt of the attack. The miners began to retreat down into the closes, chased by the police. It seemed that the police had lost any semblance of self-control, and were lashing out furiously with their batons at anyone who stood in their way. I saw one young lad struck viciously across the head. Others saw that, too, and one big miner, retreating into the safety of the close, turned and raced back to where two policemen were standing over the bleeding form of the young lad. 'You bastard! You've killed him!' Incensed with rage at seeing this obvious child struck down so senselessly, the miner struck out at the policeman, once, twice, and he went down across the body of the unconscious boy. The miner went down with him, still punching wildly. The other raced across and laid into the miner with his baton. He groaned, and slumped unconscious across the unconscious boy. Other miners had by now appeared, and so had other policemen. The struggle was short and furious, but it ended by the miners retreating again into the closes, and the first man, the avenger of the young boy, being dragged away, still unconscious, to face whatever happened to him later. The injured police were led and carried away, leaving the lad still lying on the ground, and still unconscious. He was carried into the nearest house, and the blood washed off his face and body, while a very angry crowd stood by the door, waiting to hear about his condition. He was a well-liked lad called Lance Barr, and although still at school worked in the butchers' shop at weekends, and so was known by everybody. He had not been seriously injured by that totally unprovoked attack, but he carried a large scar across his forehead for the rest of his life — a memento of which he was very proud.

There were other skirmishes that day, and for several days following, as the police tried to escort their little handfuls of blacklegs to the mines. It all ended quite suddenly when the few blacklegging families did a moonlight flit and left the area. They were never forgotten, and ever afterwards, wherever they were, carried the brand of 'Blackleg', like the mark of Cain.

The strike dragged on, from days to weeks, and then to months. There was a terrible struggle in every family to keep food on the table, and to keep the family together. There was not just poverty: that we all knew, and our parents knew how to cope with it. Now we faced real hunger and starvation. Still the men and women fought on. There were soup kitchens, and the local butcher and grocer helped as much as they could, but what they could do was little enough. At school we were given porridge, soup and cups of cocoa. This was cooked up in the schoolyard, and we thought it was a wonderful novelty. Our parents and teachers did everything they could to shield us from the worst hardship, but they could not shield us from some understanding of the misery our elders were undergoing.

After school we often followed the men down into the woods, where they would saw down a tree and cut it into lengths, and then men and women together would carry it back and cut it again into logs. Everyone got their share, and the waste wood was then gathered into a heap for a bonfire, into which potatoes were placed for roasting. I remember them being pulled out of the embers with long pieces of wire, and shared out among the children. I remember, too, how they tasted. Often enough the bonfire was also the scene for a bit of entertainment, unusual enough in those bitter-days to be recalled even today. A man called Nicol had an accordion, and he would bring it out and sit on the steps of the Buildings, near the bonfire, and play jigs and reels and the music hall songs of the day. Men and women and children gathered around, and there was some dancing and singing, and for an hour or two the utter misery and increasing hopelessness was banished from the minds of those struggling people.

The winter came on, and the men were still out. It was no longer a General Strike, of course. That had lasted only a few days, and then been betrayed by leaders — not miners' leaders — who had never sought it in the first place, and were terrified of it when it happened. The miners were alone in their struggle, and were feeling the pinch. The houses were cold and damp. There was no coal now to fuel that big constant kitchen fire which was the very heart of the household. The Pack Woman came no more: there was no money now to buy her second-hand clothes and shoes. winter was biting deepest when the men found an outcrop of coal down in the Milburn area. It was not very thick, but at least it was coal. The bings had been raked over again and again for every scrap of coal that could be recovered from them, and the fires were burning low in every house. The outcrop could be the saviour.

And so it was. Except for two young girls, who were buried in a cave-in. They had been working there, and had dug too deep. The men working the outcrop dug with desperation to get the lassies out, but it was too late. One was dead when they reached her. They carried her body home, wrapped in an old sack, without even the dignity of the carter and his old horse, and in procession went to her home and laid her down, another martyr, another unsung victim. Cha and I watched the sad little procession, and saw men and women weep as it passed.

Somehow the long months of the strike wore over. We children were sheltered from its worst effects. We were never really hungry and never really in want, although our parents suffered deep and genuine hardship. We did not really understand what was going on, but we did feel the increasing tensions and frustrations and deprivations of our parents. Finally, the men went back to work, those who were allowed to, beaten and bitter, but certainly still proud and undefeated.

Joe Corrie
Sonnet

Noo I ha'e broke my fetters and am free
   To lead the life I please, far frae the strife,
And poverty, and pain, and misery:
   I canna say I'm yet content wi' life.
Nae pit lums reekin' here, nae hiss o' steam,
   Nae risin' in the mornin' stiff and sair.
Nae slavin' in a gassy three-feet seam,
   Hungry and tied wi' hardly the breath to swear.
No, that's a' by wi' noo; rise when I like,
   The flo'ers bloom roon' my door, the mavie sings;
But where's the peace I soucht? Hungry, on strike,
   Battlin' for life — The memory to me clings,
And haunts me nicht and day. And in the pit
   My auld mates lie and slave and battle yet.

Len Doherty
Collier's Complaint

He could have said a lot more to Herbert. He should have knelt beside him, shown him his own lacerated hands, and told him:

'Listen. I said you've got to live right. That means using these right. Because a man can do anything with these. Don't you see that? You could build a world — a wonderful clean, good new world — free of this cheap useless living you hate so much. A world where nobody need feel lost — no one cried in the darkness like you and me and a whole lot more do now. Hands are just tools. But they're beautiful powerful tools — it's all in how they get used.'

They were bad, for instance, he thought grimly, when you made them become fists. It was easy to insist to others that Mathews had started it and that he — Robert — had paid enough for his own part in it. But in the end every man had to be his own judge. Mathews was dead, and he, who held human life as sacred above all else, had killed another man.

And now his mind had turned full circle and he was asking himself how many times he would have to go over all this again.

He stubbed the cigarette out and fell back on his pillow. He heard again his father's deep agonized coughing and wondered how his mother could sleep through it. Then his own problems began shifting around in his head again and he moved restlessly, with sleep a long way off. Perhaps it would be better to get away from here. People who had known him would always be looking at him in a way that forced him to remember. It was hopeless thinking you could go on for ever smothering your own thoughts.

Abruptly he sat up and threw off the bedclothes. He knew he wasn't going to sleep and he couldn't lie still with it any more. He pulled on his trousers and socks and went downstairs, stepping quietly so as not to disturb his parents. There were a few embers still smouldering in the living-room fireplace and with a sheet of newspaper he drew them into burning life and then built a cheerful fire. He washed his hands and put the kettle to boil on the gas stove and was standing in front of the fire when his father came down.

'What's up?' his father demanded. Like Robert, he was wearing only his shirt, socks and trousers.

'Just restless,' Robert answered vaguely. He looked at the ashen features and tired heavy eyes beneath his father's sparse rumpled hair and moved quickly to pull an arm-chair nearer to the fire.

'Sit here,' he said. 'Want some tea?'

His father nodded and took the seat, leaning forward with his back hunched-up to warm his trembling hands — though it wasn't at all cold. He breathed in shallow little gasps, rationing the amount of air to reach his sore lungs. Turning to Robert, he squinted up at him.

'You've got too much on your mind,' he growled. 'Where's all this reading and stuff going to get you?'

Robert shrugged. He considered telling his father all about it for a moment, but decided against it. There was no point in probing into it and digging it all out for someone else to hear, and besides — he didn't feel like it. He heard the kettle start to hiss and bubble in the kitchen and hurried out to it. By the time he had made the tea and poured out two mugs of it his father seemed to be feeling a little better. He took the mug Robert handed to him and drank from it gratefully.

'Ah,' he sighed. 'Better'n all their cough-medicine is this.'

Robert sat down in the arm-chair opposite nursing his cup in his hands. He looked at his father's white face and grunted his scepticism.

'It's more than cough-medicine you need,' he said bluntly.

'Eh?' his father became wary. 'What're you getting at?'

'Bronchitis!' Robert paused and his father waited. '— What is it?' he went on softly. 'Pneumonoconiosis?'

For a moment it looked as though the other were going to deny it. He sat there glowering with his face obstinately set, but then he suddenly shrugged and his features relaxed.

'That's the fancy name for it, I reckon. Silicosis — collier's complaint — you know what it is all right.'

'So —' He shouldn't have been shocked at all, since he'd already known, yet he sat there staring at his father completely unable to speak for the moment. Mellers seemed glad it was out now. He watched Robert with a little quizzical smile on his face and sipped his tea silently.

'How bad?'

'Well now —' his father pretended to be reckoning it up. 'The doctor says I've got three years if I don't pack 'pit up. But you know what doctors are.'

'But — Look here —' Robert put his cup down clumsily. '—You should be having proper treatment. They can help these things now. There's injections, and —' He stopped as he realized why his father was smiling in that queer way. '—You haven't tried to do anything about it,' he accused him. 'You're just letting it kill you off, aren't you?'

His father looked away.

'Think I want to spend t'rest o' my days in bed? I'm all right like this. It suits me better.'

Robert sat back and nodded to himself. So now he was proved right in all that he said before, and now he knew why his father had argued so bitterly against him in front of Herbert and Mainwaring. That wasn't much comfort, for it also proved that what he had said up to now had done nothing to alter his father's ideas. He needed some different approach here. Now, if ever, was the time to discover what his vaunted knowledge was really worth. He thought for a while and then leaned forward with his face shrewd and tensed.

'I don't suppose you've studied the old lady?' he demanded.

'Thought about her when you're too sick to move? When she's got you to nurse and worry about for months maybe. You know how it drags on — this. Thought about her having to watch you dying slowly?'

Mellers stared back indignantly.

"Course I've thought about her. That's just it. I don't want to be nobody's burden. I'd rather stick in t'pit till —' he paused, a little embarrassed because it sounded like heroics. '— Well, the longer I work, the less waiting they'll be.'

'You hope!' Robert cut in with a sneer. 'That's just your opinion — you want to think that, so you kid yourself it's true. Easy to convince yourself you're doing the right thing isn't it?' — He almost added: 'I know,' but he caught himself and, shaking his head vehemently, went on: '— If you face the facts properly you'll see clearly enough that it won't work out like that. It's a thing that puts a man on his back a long time before it kills him, usually. Why shouldn't it do that to you?'

The words were cruelly spoken and Robert felt sorry for his father, but he knew what he was doing. And he had done even better than he knew, for his words had struck a spark in his father's mind that set off the fuse of memory. The older man was suddenly once again a part of that unhappy home where another collier had lain in his chair fighting bitterly against death with a son too young to understand — where a wife had faded silently and without crying while in her heart she shared — must have shared — every moment of her husband's struggle. He remembered more about his mother; how in the years that followed she had seldom smiled and had laughed even less. He frowned, lifted a hand, and silently lowered it.

Robert saw the pain in his face. He had a moment of doubt, wondering whether he was really fit to advise anyone else, since he'd realized his failure with his own life. But this wasn't a mere matter of feelings or rights. It meant a man's life — and more — it meant all that he stood for, still believed in devoutly, against all the defeat and running away in the world. And in a way, it was all connected — all their problems hung together — not running parallel to one another, but connected up and criss-crossing in their lives, — so that to solve another's problems meant to strengthen his beliefs and to help in solving his own and therefore every solution found for one eased the burden by that much for them all.

'You've only really thought about yourself yet,' he said in a kinder voice. 'What about the others who've got a stake in this?

What about the ones that might — need you. Me? Herbert? The old lady?' It took an effort to say the words. He was fighting against inhibitions that from childhood had made him hold this kind of talk and emotion as effeminate and unmanly.

Mellers stared at him. Distaste flickered in his eyes and his features showed acute discomfort. He fidgeted in the chair and looked down at his unsteady hands. There was no trace left by now of his usual iron self-confidence.

'We've never been —' his hands gestured vaguely '— tha knows — we were never ones for fussing o'er each other. I don't see as I've been so much to you as you'd be losing a lot.' His voice held a mixture of half-regret and revolting pride, as though he didn't know which way he wanted to be answered.

'All right!' Robert got up and began to walk around, not looking at the other. His father didn't look up at him either, which made it a little easier.

'All right,' he repeated. 'So you've never been a family man. That was too sentimental for such as us. Well, who's blaming you? Maybe you never had much chance. Maybe that was the real reason. I suppose we're quite proud that we're not ones for showing our feelings or talking about them — how many of our kind do? But —' he stood behind his father's chair and put his hand on its cushioned back. '— That doesn't mean we've got less feelings than the ones who show them more. It doesn't mean we don't feel everything just as much — and love and hate and suffer — just as much as anybody. I'm tired of hiding it all.'

He straightened up and stepped back. This wasn't meant for his father alone, this that he was discovering. His own words, his own deep disturbance, was taking him beyond this one incident towards a true insight of what was much bigger — much more important.

'We're all involved,' he said slowly, knowing he could never express it properly, thinking of Herbert and the others — and himself. 'We're all part of the same kind. Nobody's just one person who can chuck himself away and think he doesn't count.' Then he stopped as he remembered it was his father he was trying to convince now. 'We'd be losing a lot,' he said simply.

His father didn't answer, didn't even look up. He sat with his face turned towards the fire and his hands mechanically rubbing against each other. Robert picked up their cups and took them out into the kitchen. Perhaps he did feel embarrassed now and maybe his father did too. But there had been something in that silence that followed his words, something between the two of them — understanding, sympathy — there wasn't an accurate definition. — But certainly it was something they'd never shared before.

He poured out some more tea and sweetened it. After all, fifty-three wasn't really old — except by the standards of men who spent their lives on the coal-face — and if he could last three years in the pit, what would fresh air and healthy living do for him ?

But he didn't try to resume the subject immediately when he took the tea in. He gave his words time to work while his father drank his tea and pondered.

'Tell me something,' he said casually when Mellers looked at him. 'Why did you take it so calmly when I told you Herbert had joined the Party?'

Mellers shrugged.

'Why not?'

'You were dead against us a week ago.'

Another enigmatic shrug made Robert raise his hands in despair.

'I can't make all this out. I've been home about a month and everything's altered. You're not going mad about the Party — Herbert wants to change the world in no time — everybody seems to have changed — except me.'

His father lifted his shoulders and half-smiled.

'Well — our Herbert's improved the past week or two. You've done him some good.'

'Me? I haven't done anything. Herbert's done it himself.'

'Aye?' his father evidently didn't believe him. 'That's not what he says, any road.'

Robert guessed that word had been coming to his father from Betty through his mother. Mellers squinted sideways at him without moving his head.

'Herbert's a bit took-up wi' you and you mates.'

'The Party-lads?'

'Aye.' His father looked very thoughtful. 'Full on it them lads, ain't they? Reckon they're going to change the world for theysens and nobody's going to stop 'em, eh?'

In the older man's reminiscing eyes, Robert could see Rodgers and his pals arguing and talking, voicing their struggle once more. And through this outsider's viewpoint, he could feel what the other was feeling too; the impression of latent developing power, innate strength, the same potential greatness that he had seen in Herbert and that was in all of them in varying degrees. It all concurred with what he had already said.

'Yes —' he nodded slowly and with deep conviction. '— They'll change it. Listen —' he waited till his father was looking at him. 'It's in the way they see life. In what they're living for. Maybe you were the same yourself once — only you lost it. Well, they won't lose it because they're insuring that they'll always be using it — and it's going to grow as they use it.' His ideas were coming to him much more clearly now.

'There's your answer too,' he went on with a quick gesture at his father. 'I know why you feel you need the pit. I said it once — remember? You live by it. I said that you'd solved things by giving yourself up to it and you'd have nothing left without it. Well, in the world these lads are after, that would be a wonderful thing to feel like that, because in their world a man will always be able to work and he'll be able to enjoy it — it won't just be something he's forced to do to live. But in this world we're living in, it's not enough. You're not getting the benefits of what you do — you're not really working for yourself and your people. When what you've worked for gets shared out, you get the leavings — not the fruits.'

His father frowned throughtfully and Robert was encouraged. At least they had got beyond the arguing stage. At least they were both trying to reach understanding. 'In this world — or in any world for that matter — you need more to live for than you've ever had. But there's so much that you could do, right now. What you need is to find another way into life — another foundation to base your living on — another way to be a part of your own kind, your mates, your family and everyone else. You're going to have to look for that — it won't just come. You've got to think about it — in fact you've got to put up a fight. It just depends on whether you can see — or be shown by lads like Rodgers — that it's worth it.'

It seemed as though there were two of him and, inside him, one struggled for supremacy and expression against the other in a fight that never ceased. So now the side that believed and had hope was in full sway and his fears and doubts could not be half so desperate as he had imagined. He was really alive now, confident in himself — holding the reins and enjoying the exhilaration that belonged to these moments. Yet at the same time was whispering the conviction that this was merely another temporary victory, that if he relaxed for a moment he might lose all this and therefore he must not relax.

'You're the one for fighting,' his father murmured, watching him half-admiringly. 'You and your mates.' He drained the tea and put down his mug, nodding to himself. Then he looked up.

'Well — I'll admit as you were right o'er some o' the things you said, lad. But I never denied you'd got the brains.' He seemed to fall into a frowning dream for a few moments. Then he hunched up his great shoulders again and shook his head slowly.

'It makes a man feel old,' he said in a low voice. 'Seeing his lads wi' all their ideas and running around — all on the go. I seen some action lately any road. Even your ma,' he smiled to himself.

'Well, the rest'll be worth seeing too,' Robert asserted eagerly. 'It will! It'll be worth living to see, believe me. Because it's only started around here. Wait till we really get going.'

The 'we' was a direct product of all he had been thinking. It was the first time that he had seriously counted himself as being a part of this group — subordinate to it — and one of the lads. The change was partly unconscious but he knew that he did feel a lot better. And yet this was what he had wanted when he first came home and there was only himself to blame that it hadn't been that way. Then what had stopped him?

He was unable to answer that because it was too soon for him yet. Whatever it was he was only just in the process of emerging from it and still too much involved in it to get it into perspective or be objective about it.

While Robert was thinking about it, his father stood up and, with a yawn and a stretch, decided he would go back to bed now.

'I'll see you in t'morning,' he said in his old, gruff way. 'Mind and turn them lights out, now.'

Robert nodded and said goodnight. He was quite satisfied with the talk they'd had and there would be no sense in pushing things yet. His father would be thinking about it all again now, and that was enough. It was like everything else, anyway. The final decision had to come from within.

For a long time he sat on alone, while the fire burned low, and thought of a lot of things that had been obscure lately. One thing was becoming apparent to him out of the discussion — something he could sum up into a lesson for himself. The only way that he was going to succeed in living with himself would be in learning to live with others again.

Idris Davies
Man Alive, What a Belly You've Got!

Man alive, what a belly you've got!
You'll take all the serge in my little shop.
Stand still for a minute, now, and I'll get your waist.

Man alive, what a belly you've got!
Oh, I know it's only a striker's pay you get,
But don't misunderstand me, Hywel bach;
I depend for my bread on working men
And I am only a working man myself
Just Shinkin Rees the little tailor,
Proud of my work and the people I serve;
And I wouldn't deny you a suit for all the gold in all
    the world.
Just pay me a little each week, Hywel bach,
And I am your tailor as long as you live,
Shinkin Rees your friend and your tailor,
Proud to serve you, and your dear old father before you.
But man alive, what a belly you've got!

Dennis Potter
The Revised Plan for Coal

When what prosperity there was, and even the situation and
colouring of the sprawling, hilly villages and little towns of the
Forest, had been decided mainly by the local mines, the few, bleak
words in the *Revised Plan for Coal* came hard and painful to the
people there. At least, it was assumed, their utter inadequacy
might have been filled out with the platitudes expected of those
who rule the lives of the miners.

I talked with many of the men who worked in the two pits
nearest to Berry Hill, Cannop and Waterloo, most of whom I have
known all my life. My father also worked at Cannop for nearly
thirty years after he left Waterloo. At one time, practically every
home in the village had a member of its household (and sometimes
three generations) working in one or both of these two pits.

A streak of self-pity inevitably affected the way they talked about
what had happened to the local coalfield. 'It was alright when we
were wanted. National bloody heroes. Nothing too good, jokes
about us on the wireless.' They realise that the N.C.B. carried on
operations in what was fast becoming an uneconomic area —
through exhaustion and dwindling seams — only because of the
country's urgent need for coal in the years of reconstruction after
the war. They remember that here, too, in a once poverty-stricken
area, besides the 'guarantees' of the recruitment posters, came the
insistent plea to the men to work a six-day week of full shifts and

as much overtime as possible. This, it is acknowledged, would have been perfectly okay if, while time and opportunity had existed, a smoother, more human, and efficient run-down had been planned. But, as Oliver Oakey, a recently retired manager of Cannop told me, 'always they were on to us — coal, coal and more coal. It didn't seem to matter about anything else. Just coal — and we couldn't get enough for them, couldn't do enough, and no matter how much it cost per ton.' So the Forest, as always, had an active life only according to the state of the market for coal; good times alternated with the bad for this reason only. As a result, there is a great deal of scepticism about politics, including Labour politics, amongst the older people of the Forest, and the usual indifference of the young. And statements couched in high-flown and rhetorical idealism are greeted with scorn, with not even the polite shield of lip-service.

Through all the early days of the war, when work came back and the men responded to the slogans chalked on their shovels, and through the years of 'Export or Die' and the desperate winter of 1947, years when the miners were treated as heroes and medallioned giants, the decline of the Forest as a coal-producing centre (though scarcely whispered about at the time) has been 'as sure as God made little apples', as Jack Hawkins put it. Yet of 458 miners who lost their jobs with the closure of Eastern Untied, only 79 of these were promised new jobs by the Coal Board. No plans were made for the rest, and, since they did not want anyone to leave before necessary, as little notice as possible was given. Hence the wreath of holly, and the queue at the Labour Exchange.

And yet a place that thinks a lot of itself, that has a sense of its own identity, is frequently able to make a nonsense of jaded descriptions of alleged apathy. There is, as a result, an astonishing air of liveliness about local councils and unions in the Forest of Dean, an illustration of a kind of reality about local democracy and concern that I had been inclined to doubt as a possibility. The district councils, tradespeople and unions formed the Forest of Dean Development Association in an effort to counter the too dangerous, too familiar limitations of long dependence upon a major industry with such a fluctuating record as that of coal-mining. The Development Association is not simply a paper organization, or a body working in isolation and without publicity or local concern, but something which has managed to be talked about, written about, argued about, praised and condemned with a vigour which shows how much it has been able to get a frustrated quietism turned into a rebellious and muscular concern about the continued life of the Forest as a place where it is nice to live and bring up children. Most of the new factories in Dean have come about this way, with headlines of triumph in the local newspaper, motions of congratulation on the local councils and protestations of support and enthusiasm from the local trade unions. Part of this unanimity and energy may come from the fact that, by and large, the Forest is still a one-class area with a common accent and an extremely powerful, almost chauvinistic sense of its own values and traditions — 'our Forest humour', 'in this little Forest of ours', 'it's a little country on its own', 'once a Forester always a Forester'.

'If we don't help ourselves,' I was told, and it was agreed, 'nobody else will.' And while the Coal Board, for its part, issues a soothing but meaningless statement to say that 'the Divisional Board are deeply concerned about the position in the Forest coalfield', the miners continue to think they have been forgotten, made use of in the past, and then dumped like the unsaleable coal itself. They sense the confusion and lack of imagination that has settled in 'official' minds; for co-operation between the National Coal Board and the Board of Trade at a national level, and between the groups and Divisional Boards and local authorities nearer to home, has been and is far too shadowy to offer any real hopes of the planned resurrection for the area which might have taken place. The miners that are left, skilled through a lifetime at their own trade and ignorant of any other, are all too certain that the withdrawing Coal Board will be in no position to train the men, the ageing men, it is leaving behind; for, is it not obvious, coal miners have not given the same service to the community and are not as worthy of help as redundant and pensionable army officers.

Two summers ago, my father got out of Cannop, sick of a pit that had become swamped by dissatisfaction, rumour and counter-rumour, and managed to get a job at greatly reduced pay and much longer hours as a cleaner in the Red and White bus garage at Coleford. He was, comparatively speaking, fortunate to have been so placed, because he is in his fifties and avoided some of the rush of men of his age and experience which has weighed down the Forest over the last four years. But he finds his present job boring and irksome in a way that, as far as he is concerned, mining could never be. Pushing a broom along the ridges between the bus seats, he describes it, where once he had listened for the slightest creak, the ache in the timber long preceding any move-ment in the roof above the stall: mining certainly engages a man's attention, and gave moments of pride, while the hate for it was not the contempt he is now enduring in his new job.

Mining was possessive, and bred its own ethic. The ex-miners in the Forest of Dean miss this, and find it difficult to pin down their exact objections to factory life or other work. 'The trouble is, you know,' I was told by the Personal Assistant at a local factory, 'these miners are difficult chaps to employ. Always ready to make trouble, and they find it difficult to settle down to factory life.' Brown overalls for those on maintenance, 9 a.m. for staff, 7.00 or 7.30 or a shift system for 'the workers', a divided canteen, a factory magazine which is, of course, an employers' magazine, a weakening of Trade Unionism and the atmosphere of individual promotion, the faint possibility of individual bargaining — all this is felt to be deeply obnoxious, a trap. The miner does not believe in 'emancipation' through individual grace or annual bonus which is a percentage of annual wage. The younger Forester, however, is not altogether so sure about 'our side' and 'their side', about 'us' and 'them', and is not so conscious of his bargaining-power either. There has been a considerable decline in the political consciousness of the area, as shown in the flavour of the talk and in the faint but discernible mental shrug with which the older concepts of 'working-class' ambition and purpose are met. 'It's no use,' I was told by an acquaintance from Mile End, 'I can't talk to our old man — he's always on about this or that, about what we're supposed to be cheated of, that I just give up.'

Meanwhile, those who still work in the pits, mostly middle-aged men now, continue to squat on the village crossroads to wait for the workmen's bus which takes them through the trees to the mine. At Berry Hill, a few wait on the corner opposite 'The Globe' with its Ansell's Beers sign and yellow-wash walls, and the signpost which says 'Joyford ^{3}/4'. They wear mufflers and caps, and there is that washed-out paleness on the older faces, the blue coal scars, and 'bread' in old Home Guard or army satchels and Oxo tins. A way of dressing, the kind of stance, has already begun to date them: the young Foresters do not squat on the backs of their heels as these do whenever they have to wait. These miners are some of the men who laid the huge and attractive Berry Hill rugby football and cricket ground during the 1926 strike, whose minds are full of memories and talk of rugby, the band or the chapel or Coleford's Fair Day. But they are not posing for a photograph, for they themselves have willingly changed, or drifted along with it like everyone else. Behind them, Hawkins' Stores, now a limited company, is building a new extension, serving the young families in new council houses up the road, offering Danish blue cheese, telly snacks, striped toothpaste and coloured toilet rolls.

For a moment, if you did not visit them at home or talk of other than traditional subjects, the men outside the shop might appear to wait like monuments, carved out of the thick substance of the idea of a miner, the idea of the twenties and thirties, so that they will seem, almost, to be statements that have been completed, expressions that are no longer relevant or certain. A charabanc on the way to Symond's Yat Rock, jutting out above a loop in the Wye, whirrs past the mediocrity of the village; but the visitors inside it, heads back against the padded top bulge of the bus seats, stare for a moment at the groups of waiting miners, place them in the newspapered recesses of their minds, and wait for 'the scenery' to reappear once more. 'Hold it!' a voice seems to shout. Snap. And we are safe again with Andy Capp, or 'working-class' poses, or easy political claptrap from right and left angles.

Harri Webb
Butties All

We don't bear Churchill no malice
Nor think of his name with ire,
Every night in Blenheim Palace
He tubbed in front of the fire.
He cussed and swore something awful
Like all the colliers do,
He died of M.P's silicosis
And all his scars were true-blue.

Bob Smith
1974

It became clear that another strike was approaching. That was in 1974, and again the issue was wages, but this time there was also the very serious issue of a national plan for coal, which we, rightly, saw as determining the very future of our industry, as well as our own personal futures. And we were determined to protect them. Obviously it was going to be a bitter struggle, and we prepared for it. Strike committees were set up in local areas where we could contact them in the event of any emergencies or for arranging picket lines. We had two focal points in our district, the two power stations. We placed a caravan at each of those sites as shelter and headquarters for the pickets. It was my job to keep those caravans supplied and provisioned. There were also many meetings to attend and much correspondence to deal with. A lot of that correspondence was concerned with keeping the picket lines quiet and peaceful and lawful, and I stressed the importance of that at meeting after meeting. Peaceful picketing is never easy, and tempers can flare quickly at times, but that is what we stressed at every opportunity.

On one occasion we had a picket at Kincardine Power Station. There were several of our men and two policemen, one of them a sergeant, and we chatted and joked to pass the time. Several lorries had driven up, and we had talked to the drivers and they had all turned round and gone away, refusing to cross our picket line. We cheered each driver as he drove away. Another lorry came up and we spoke to the driver. He also agreed to go back, and we gave him a cheer as he went. However, a couple of minutes later he came roaring back up the road and crashed straight though our picket line, scattering pickets and policemen as we all leaped to safety. Two pickets and a policeman were slightly injured, and the lorry went on to smash into a car containing some reporters, and wrecked it, before hurtling on into the power station. Some of the lads, screaming abuse and threats, raced after it as far as the closed gates. Others assisted the reporters from their wrecked car. They were only bruised and shaken, so we took them and the policemen back to our caravan for tea and food. We dressed the wounds and stopped the bleeding, and gave them all sympathy.

There were fearful threats uttered about what would happen to that driver when he came back. Wullie MacDougall, who was there also, and who had leaped into the hedge together with the police sergeant, had his work cut out to calm everyone down. He had this gift of taking the heat out of a difficult situation, and he did it again. We discussed the situation with the police. We were obeying all the rules and laws of picketing, onerous and unfair as they were, and in no way could we condone the action of that driver. He had been reckless and dangerous. He had wrecked a car and injured its passengers. We would have liked to have got our hands on him and I suspect the police would have, too. However, justice, if that is the name for it, took its course. The man appeared in court, and was duly admonished. *Admonished! For what he did!* Some of our lads went to jail for even being on a public road leading to a picket line!

The strike went on, and our situation grew worse. There was no money coming into the house, except a pitiful DHSS allowance for my wife and younger child. It was a real struggle to keep afloat. My cigarettes were the first thing to disappear and I forgot the taste of a pint. As Branch Officials we were continually on the move, with picket duty, soup kitchens and meetings to keep the men informed. Wherever we went, we encountered the same closeness and comradeship and determination.

It was one of my jobs to keep our two caravans supplied with provisions, and I was amazed at the amount of tea consumed. I commented on this and one of the lads asked if I had heard of the saying that when the people had no bread, they should eat cake. Well, we had no beer, so we drank tea!

There was one morning when I arrived at Kincardine Power Station about four o'clock. I expected the lads to be drowsy and half asleep. Far from that — they were all there, all puzzled and alert. Mick Fox greeted me. 'Did you ever see the likes of this? Kincardine is full of police. The police are everywhere.' I told him not to worry, and together we walked down through the village. He was right. The police were indeed everywhere. There were panda cars and uniformed men standing in groups. They eyed us up and down as we walked along. It seemed like the British Army in blue uniforms. All the phone boxes were out of order, having been 'spiked', obviously to prevent us from contacting the other strike centres. We went back to the caravan and I arranged to get transport immediately over to Fallin, the other power station, to report the situation and get advice. I reassured our men on the picket, and told them they had to stay cool.

Arriving at Fallin I found Terry McMeel there. He was the leader of that section, and he laughed when I told him of the situation at Kincardine. He said it was all a hoax. Some bloody hoax, I told him, and reported how the village was full of police-men and of how the phone boxes had been put out of order. He explained that the previous night someone had rung up George Bolton, and asked about the situation at Kincardine. George said, jokingly, that a mass picket would be there, and that the power station would be stopped.

Anyway, there was no trouble at Kincardine that day, although there was a good deal of provocation, and I had to work hard keeping our lads cool.

The hours were long, and it seemed that I had not spent time with my family for weeks. Whenever there was a break from picket duty, there was a meeting to attend or correspondence to deal with. May was her usual kind and understanding self, and fully supported the strike, and was prepared to suffer along with the rest of us. The difficulty was the children. Not that they went short of anything essential, but everything had to be rationed carefully, and the one thing I wanted to give them above all else, my time, I had to give elsewhere.

So far as I was concerned, trouble first came when I was on picket duty at Longannet. When we arrived to relieve the night-shift, we found the police were already there in force. We were planning a bigger turnout ourselves that morning, so already the potentiality for trouble was present. More police arrived, and marched down the centre of the road, with our men lining both sides, and whistling Colonel Bogey as the police marched along. It was all good natured, and the police seemed to be laughing as much as we were.

The police then lined the road leading to the power station gates, and our lads were behind them. There was a bit of pushing, but it was all perfectly good humoured, and jokes and patter flew. Following the established procedure, each car and lorry was approached by two pickets, who attempted to persuade the driver to turn back. Sometimes they succeeded, and sometimes they failed. Each success was greeted with cheers, and each failure with boos. One lorry driver charged straight through the line, causing one picket and one policeman to leap for safety. There was a howl of anger from the pickets and they strained forward against the line of police. If the road could be closed completely, it would not be necessary to argue with each driver separately, and run the risk of being run down by some stupid person.

There were shouts of 'Heave! Heave!', and the police line wavered, swayed, and was broken. The men poured into the road and blocked it. There was a bit of pushing and shoving as the police tried to get the road clear again, but it was mainly quite good natured and joking. Eventually the lads did return to their lines, and again there was peace, and we got on with our legitimate picketing. Then a heavy lorry escorted by a police car thundered straight through the line, despite all the efforts of the pickets to stop it. Again there came the call to 'Heave! Heave!' and again the police line was broken, and pickets streamed across the road. Tempers were beginning to flare by now, with the police irritated by their inability to hold their line. There were some scuffles and some kicking and punching by the police as they fought to get the pickets back off the road.

Our lads objected to this, and very quickly the atmosphere changed from one of reasonable good humour to bad temper and conflict. It was ugly, as pickets met police head on and battled with them.

Several arrests were made, and the Union Officials did their best to restore order, appealing to the men to stay cool and return to their ranks. Reluctantly, they did, but the arrests continued and this inflamed the situation, especially as those arrested were being beaten up in full view of their fellow miners. Graham Steel, a member of the Executive, appealed through a megaphone for restraint, and for a time there was restraint, on our side. But again the arrests went on, and the public beatings. Still, our lads were disciplined enough to recognise provocation, and refused to respond. The road was again cleared, and the lines restored, and the pickets on duty again began stopping vehicles. I was one of them, and enjoyed arguing with the drivers, especially when they told us of their sympathy with our struggle, as many did.

As I was relieved of duty and was heading back for the caravan for the tea I felt I deserved, there was a heave from the pickets, and a group of men, perhaps half a dozen, spilled out into the roadway, probably in fact pushed from behind. They were just by me, and suddenly we found ourselves surrounded by a group, a gang of police, who punched and kicked at us. That was no place to be, so I slipped between two policemen, and headed back for the lines, stopping a few punches as I went, but very carefully not retaliating. I got free of the struggling bunch, and then suddenly my arms were twisted up behind my back, very painfully, and a voice grated: 'We've got one of the bastards!'

'Hold it,' I called, 'I'm only heading back for our own line.' 'That's what you think!' replied the burly young policeman who was busy giving my twisted arms another heave further up my back. I was frogmarched away, despite my protests. 'We are having you, you little bastard,' the policeman snarled. 'We've had enough of your fucking kind!' I was held helpless, and beaten about the arms and legs, and then pushed over towards the police van.

Some of the pickets saw what was happening, and they roared with rage. They heaved forward, the police line broke and our lads raced forward to where I was being half pushed, half beaten into the van. The two policemen holding me saw what was happening, and realised what was likely to happen to them in a moment. They gave one last jerk to my arms, then pushed me forward into the road and had one last kick at me. I felt a searing pain in my ribs as I staggered to my feet and stretched out a hand for assistance from someone. A big miner pushed his way through our now thin ranks and put his arm around my shoulder. He led me off and sat me down on a rock. Then a hand grasped me and pulled me up and round. It was a police inspector, and he was angry.

He was actually snarling, and his lips were really twisted with hate. 'You,' he told me, 'You get up that fucking road and if I see your fucking face down here again you will be in the clink.' I said nothing, and did nothing. He jerked at my arm. 'I know you,' he said, 'I know who you are and what you are. Now get to hell up that road or I'll have you in the van along with your fucking mates!' I still made no reply, but gently disengaged my arm and walked into our lines. He glared, and then walked off into the police cordon.

Back in the caravan we talked over the situation, and I was given tea, the usual remedy for all ills. It was difficult for me to breathe, and clearly I could do little more on the picket. Anyway, I had been there since early morning, and it was now late. Time to be away, so a car was organised for me and a few others who had been hurt. I was quiet on the way, and the others, who knew that I had been beaten up, told me that it was time I had a rest. I looked white and weary, and needed to take some time off. I had to agree that I was weary, and agreed that I would hand over for a while to others, and get some rest. At our strike headquarters I reported what had happened, and then told them I was off for a while. Everybody agreed, so I went home, to the care and love of May.

Idris Davies
Let's Go to Barry Island, Maggie Fach

Let's go to Barry Island, Maggie fach,
And give all the kids one day by the sea,
And sherbet and buns and paper hats,
And a rattling ride on the Figure Eight;
We'll have tea on the sands, and rides on the donkeys,
And sit in the evening with the folk of Cwm Rhondda,
Singing the sweet old hymns of Pantycelyn
When the sun goes down beyond the rocky islands.
Come on, Maggie fach, or the train will be gone
Then the kids will be howling at home all day,
Sticky with dirt and gooseberry jam.
Leave the washing alone for today, Maggie fach,
And put on your best and come out to the sun
And down to the holiday sea.

We'll carry the sandwiches in a big brown bag
And leave our troubles behind for a day
With the chickens and the big black tips
And the rival soup-kitchens, quarrelling like hell.
Come, Maggie fach, with a rose on your breast
And an old Welsh tune on your little red lips,
And we'll all sing together in the Cardiff train
Down to the holiday sea.

Duncan Bush
Summer 1984

Summer of strike and drought,
of miners' pickets standing on blond verges,
of food parcels and

hosepipe bans.... And as (or so
the newspapers reported it) five rainless
months somewhere disclosed

an archaeology of long-evicted
dwellings on a valley-floor, the reservoir
which drowned them

having slowly shrunk towards
a pond between crazed banks, the silted
houses still erect,

even, apparently, a dusty
bridge of stone you might still walk
across revealed intact

in that dry air, a thing not seen
for years; just so (though this the papers
did not say)

the weeks and months of strike saw
slowly and concurrently emerge in shabby
river-valleys in South Wales

— in Yorkshire too, and Durham,
Kent and Ayrshire — villages no longer
aggregates of dwellings

privatised by television, but
communities again, the rented videos and tapes
back in the shop,

fridge-freezers going back
— so little to put in them, anyway — and
meetings, meetings in their place,

in workmen's clubs and miners' welfare
halls, just as it had been once, communities
beleaguered but the closer,

the intenser for it, with resources
now distributed to need, and organised to last,
the dancefloors stacked

with foodstuffs like a dockside, as if
an atavistic common memory, an inheritance
perhaps long thought romantic,

like the old men's proud and bitter
tales of 1926, was now being learnt again,
in grandchildren and

great-grandchildren of their bloodline:
a defiance and a unity which even sixty years
of almost being discounted never broke.

Barry Hines
Your Dad Was Brilliant

As Cal drove along, he switched on the tape recorder lying on the
seat beside him and listened to Helen Woofitt's version of his
father's arrest.

'... I mean I wasn't even involved, not at first anyway. I'd just
gone to the supermarket with our Rachel when I saw this crowd
outside his house. There was no trouble, people were just standing
around. One or two were shouting scab and that, but it wasn't nasty.
I mean they weren't going to burn his house down or anything.

'Then your dad arrived and walked up to the house to see if he
could have a word with Phil Walton, but the bobby wouldn't let
him. There was no argument or raised voices or anything between

them — in fact you could see the bobby was embarrassed. He was from round here. I've forgotten his name now, but he didn't like what was going on either. Not all the police were bastards. It was mainly them from other parts of the country that were the worst. There was no comeback for them. The local police had to live with it when it was all over.

'Anyway, we were stood around talking when suddenly this police van draws up and all these coppers jumped out and started ordering everybody about and that. That.... What do they call him? Hang on a minute, it'll come back ... Taylor ... Tolson! That's it. Inspector Tolson. He was a right clever bastard, him. He was just looking for trouble. Come on! Move on! he starts shouting, or you'll be arrested. And then the coppers started pushing us around. I mean, it gets your back up that, doesn't it? Especially when you're not doing owt. It wasn't that there was a riot going on or anything.

'Anyway, as you can imagine, the whole thing turned nasty. Scuffles broke out and they started making arrests. Our Rachel started crying. But I'd got my rag out by then. I know I shouldn't have, but I wasn't having anybody pushing me around, especially when I hadn't done owt. And the things they called us! I'll not repeat them, but you can imagine. I just told them to eff off. Fight fire with fire, that's my motto. Our Rachel was terrified by this time. She was screaming and hanging on to my legs. She was only three at the time. She was in a right state.

'Anyway, your dad comes down the path from the house and tries to calm things down, telling us not to retaliate and to keep calm and that. A fat lot of good it did him. He just got truncheoned and thrown in the van with the rest of us. The worst thing though was — and I'll never forgive them for this — our Rachel was hanging on to me for dear life. She could see they were going to take me away. But instead of letting her go with me, they dragged her off and left her in the street. When the van drove off, I could see her through the window running after the van, her little face terrified. I mean.... Sorry about this. It's all coming back....' 'Take your time, Helen....' (his own voice on the tape).... 'What sort of people are they who can do a thing like that? But they did. And they would have done anything. Anything to break the strike, because they were under orders from the government to.

'Your dad was brilliant during the strike, Karl, the way he went round people's houses giving advice when they were in trouble. I mean people were getting into terrible debt. They didn't know how to cope with it. They'd never been in that position before.

We had our electricity cut off. I'll never forget Graham's face —
that's my husband. I'll never forget his face when he came home
and I was sat in the dark with a candle lit. He'd been kicked to
pieces on the picket line all day, then when he comes home the
electricity's been cut off. I thought he was going to burst into tears,
poor sod. The trouble was, we'd only got an electric cooker so we
finished up having to cook on the fire. Not that there was much
cooking to do, mind. I don't know how we'd have managed
without the food kitchen.

'Anyway, your dad came round and advised us to go round to
the Electricity Board and claim special circumstances because of
the baby and that. He said he'd bring our case up on the commit-
tee and see if he could get us anything out of the strike fund. It
was horrible. It was like living on charity. The only consolation
was that everybody was living in the same boat. People got behind
with their mortgages. Televisions had to go back. Furniture. Cars.
Everything. I can remember Alec Todd next door. He got a snotty
letter from the bank about missing a repayment on his car. So he
picks up the phone and says, Is this the listening bank? Right.
Well, listen to this. If I get any more letters like this one, you're
getting eff all. And he slams the phone down. Joyce Grayson used
to put all her bills into a hat and draw one out. She said it was the
only way she could decide which one to pay. It was heartbreaking.
You'd worked hard to build a decent life, then suddenly it was all
taken away from you. But we stuck it out because it was a just
cause, Karl. And I said to your dad when he called round, They'll
not force him back to work, Harry. I'll burn the furniture first to
keep warm.'

Tom Earley
Rebel's Progress

When idle in a poor Welsh mining valley,
Dissatisfied with two pounds five a week,
I got invited to a marxist rally
And found to my amazement I could speak.

I soon could spout about the proletariat,
The bourgeoisie and strikes and lockouts too,
Could run an AGM or commissariat
As well as boss-class secretaries do.

At first I joined Aneurin Bevan's party
But soon got disillusioned with all that.
Joined Harry Pollitt and became a commy.
They turned down all my pacifism flat.

The hungry thirties found me hunger marching
To squat with Hannington inside the Ritz.
Then PPU. For just this I'd been searching
Before the war and long before the blitz.

I liked the people in the Peace Pledge meeting
But found that they were holier than me,
So marched with Collins and quite soon was greeting
My former comrades in the CND.

To sit with Russell next became my hobby,
Vanessa Redgrave's fame I hoped to share,
Got thrown around in Whitehall by a bobby
And then a broken arm in Grosvenor Square.

So now I'll leave the politics to others
And not be an outsider any more.
I'll go back to the valley, to my mother's,
And never set my foot outside the door.

Except to go to chapel on Bryn Sion
And maybe join the Cwmbach male voice choir,
I'll sit at home and watch the television
And talk about the rugby by the fire.

Anon.
The Pitman's Union

Now let the colliers' hearts be glad
    While plenty round them shines,
And blest contentment flows along
    The banks of Wear and Tyne.
Brave Hepburn and our delegates,
    Like rays of virtue shine,
Their fame shall long be echoed round
    The banks of Wear and Tyne.

On Bolden Fell our flags shall wave,
  Like victory's wreaths entwine,
But peace shall be the motto still,
  With lads of Wear and Tyne.
We envy not the rich and great,
  Whose dazzling greatness shine;
While we the hardy sons of toil,
  Can labour in the mine.
Our happy wives and children now,
  All former cares resign,
And sing with joyful mirth and glee,
  The lads of Wear and Tyne.
May he, who rides upon the storm,
  Protect with care divine,
From all the dangers that surround
  The lads of Wear and Tyne.
Here's health unto the King,
  Likewise the Queen sublime,
Who gave the pitmen their applause,
  That dwell on Wear and Tyne.
Now to conclude and make an end,
  May luck around them twine;
O, bless the happy collier lads,
  On both the Wear and Tyne.

Chorus.
Still round our banners we will stand,
  In love and truth combine,
And children yet unborn shall sing,
  The lads of Wear and Tyne.

Mike Jenkins
Industrial Museum

Hello and welcome to our industrial museum.

On your right there's a slag-heap reclaimed ...
a hill ... another slag-heap ...
that one shaped as a landing-pad
for bird-like hang-gliders.
Notice the pit-wheels perfectly preserved

where you can buy mementoes
of the Big Strike and eat authentic cawl
at an austere soup-kitchen.

There mummified miners cough and spit
at the press of a button
and you can try their lungs on
to a tape-recording of Idris Davies' poems.

That rubble was a 19th century chapel,
that pile of bricks an industrial estate.
The terraced houses all adorned
in red, white and blue as if royalty were visiting.

See how quaint the wax models
of women are, as they bow in homage
to polished doorsteps, the stuffed sheep
at the roadside give off a genuine odour.

The graveyards have been covered over
and lounge-chairs provided for viewing
gravestones which tell of deaths from cholera
or pit explosions. I recommend their cafeterias.

In the ruins of the Town Hall the council
give public performances, meeting
to discuss the valley's future:
their hwyl is high and hiraeth higher.

Finally, let's visit the Foot Arms
(in memory of a long-gone leader)
and listen to the last Valley's character
who lives here, courtesy of the Welsh Office, in a tin bath.

And Gold On My Neck the Sun

Len Doherty
Locating Your Body

Blackness. Silent blackness. This was his return to life. He could feel nothing and sense nothing; nor did he know who he was or where he was or what had happened to him. He was a man, he knew that. And he had a body and limbs — but he could not locate them when he tried.

It's happened before, he thought, recalling that sometimes he had suddenly wakened in the night and found himself cut off like this, unable to move a muscle or open an eye. At such times, even with eyes closed, he was conscious of his surroundings: a shadowed bedroom, the pressure of someone inert beside him and the curve of waving curtains arrested in mid-motion, as though he and the world were caught between seconds of real time in a timeless immobility. And at such times panic always struck through him and he would immediately feel the increased pumping of his heart. This one contact with reality would be enough to release him and let him be integrate again.

But this time his perceptions cowered behind the enclosing skull and he could envisage nothing.

Nothing. Why?

His lungs must still swell and his heart must still beat: he had only to feel them. But his mind baulked at the attempt, and instead of panic he was aware only of bewilderment.

For a long, long, time he was lost, like a body submerged in deep water, yet he was bodiless like a floating spark. Then gradually his self-awareness began to grow stronger, and he discovered emotion lurking under his thoughts. There was a troubled grief in him, though there was no hint of its cause, and behind that grief were more feelings too indistinct yet to recognise. Trying to examine them he felt a quick flash of fear — the dread of something awful and unbearable having happened to him — but it was gone before he could grasp it.

This isn't like it's been before, he thought, and the fear leaped in him again. It wasn't fear of this loss of identity, but of something outside. It was a primitive kind of fear, and each time he tried to concentrate on it, his mind, like a caged squirrel, started grimacing, chattering and dancing around, desperate to escape. A thought flickered:

This'll kill us —

He was suddenly terrified. He tried to stop thinking, but the right thread had been pulled and his memory began to unravel.

An appalling impact of sound there had been, and one glimpse, up and backward over a naked shoulder, of the squealing props twisting and spinning out from under the buckling steel bars. He had heard Davy begin a yell and seen him lunging away beneath the descending, expanding avalanche, and he had thought:

This'll kill us —

And now he had located his body. He lay on his side, his left leg and arm twisted and pinioned under him, his head stretched back on curving neck. Weight clutched him like a giant stone hand. He screamed inside himself, then agony and horror smothered his mind and he fainted again.

George Barker
Miners Above Ground

Dead men and miners go underground.
Deeper than vegetables or the rock,
Than the Cro-Magnon arrowhead or sounding
Whale, deeper and darker than a black
Burial, they both go down into dirt.
But the dead stay down. We forget them.
The sometimes smiling miner of Glynneath
He comes up murky as his shirt
Out of the belly of South Wales. Let them
Elated this Saturday be happy beneath
An unfalling bright sky. Their work is done,
Rigging a drift, riding a spake,
Hacking the seam. A week's work's done
And — fine and unlikely as a birthday cake —
These men enter the Saturday of the sun.

Harold Heslop
The Wicked Practice

A mine is easily ventilated. Two shafts operate as the chief factors in the scheme — the downcast shaft, down which the fresh air rushes and the upcast shaft, up which the foul air is drawn. This system entails among other things a powerful suction fan capable of keeping the air circulating through the workings, and an

effective system of stoppings and trap-doors to prevent a prema-
ture rush of fresh air direct into the return before it has completed
the ventilatory circuit. It follows, then, that the pure air is continually rushing into the workings by one route, and, having
become foul, is continually rushing to the fan by another; and it
also follows that while every effort is made to prevent direct access
from the intake to the return full efficiency is by no means obtained.
Therefore, the points of escape must be reduced to a minimum. In
the final moment the intake must blend into the return at the face.

Correct ventilation of a north-country mine still employing the
early nineteenth-century system of coal-production is a thing
honoured in the breach. The difficulty often arises in the instance
of the intake coming to the inevitable cross-roads, that is, where
the workings are diffused over a wide area. In such cases an
elaborate system of carrying the return airway *over* the intake
airway and hermetically sealing the one from the other is necessary, but this generally proves to be an expense frowned upon by
managers. The wicked old practice of making one engine plane
the intake and forcing into the return system all the others is often
indulged in. This forces the air over a wider area than is actually
permissible. It is a wicked practice, for it must be admitted that
in a mine accurate ventilation is the thing most desired. When
there is an inflow of pure air, travelling at a few feet per minute,
no man need worry unduly about explosions. It is the lack of this
healthy, invigorating flow of air which breeds the conditions for
death-dealing explosions.

Mine ventilation, while simple when viewed as a whole, demands a few simple tools — wooden doors, brattice cloth and
brattice nails. In addition, patience and foresight.

An accumulation of inflammable gas is dangerous only when it
is stationary. Once it is forced into the ventilatory seam it ceases
to be a menace. A stationary pocket of gas is one outside of the
stream. It would be a matter of simple logic to put down all
explosions as being due to stationary accumulations of gas. Gas
accumulations can go unnoticed. Unused places can be and are
withdrawn from the ventilatory current and these can be forgotten
and unexamined. Humanity does not demand such a colossal
waste of time in the careful examination of all waste places beneath
the earth but the mining people do demand a good number of
return airway men to give the main places a thorough examination. Again, a place can be boring forward well out of the reach of
the main supply of air and find an outlet of gas. There are many
reasons why these frightfully dangerous pockets of gas are formed.

'Blowers' often break through the coal from the surrounding strata and these must always be carefully dealt with by the miners themselves. No miner is criminal enough to jeopardize his mate's life by neglecting these menaces as they appear.

In the case of Franton Busty the ventilation process had depreciated so startlingly that the continuous source of extraneous gas had never been adequately dealt with. As the area had widened at these points the strain on the fan at the top of the upcast shaft had been too great and the ventilation system had almost collapsed. Even the introduction of a fan inside the pit had not contributed materially to the improvement of the situation. It was possible to get a good-sized cap of gas on any safety lamp in the district.

John Tate and Walter Oxley had arrived at the North winnings flat.

Very quickly Oxley took a reading on an instrument which an overman carried. He spoke slowly the reading which Tate jotted down in his pocket-book.

Tate nodded. 'Very humid, I must say,' he said.

Oxley did not say anything.

'Hide your lamps,' Tate commanded.

Work ceased for a few moments while he carried his lamp carefully to the roof above his head. Very gently he lowered his light until the cone of blue flame unmistakably appeared over the original cap of the flame. They were in the unmistakable presence of gas. It was not an oil cap on his lamp! And this was actually taking place yards away from the actual workings.

'Hum!' he said.

He did not hide his dissatisfaction.

'Carry on!' he called.

The work of securing coal for the earth above was resumed.

John Tate sat considering the problem thus presented.

'All right, Oxley,' he said. 'You can go.'

Walter Oxley and the overman went away, as had been arranged, and left John Tate sitting out of the way pondering this problem. Certainly the place had become sadly neglected. The ventilation had been sadly neglected. It ought to have been seen to earlier than this, and he knew that had he spent some time, money and energy upon putting this place to rights instead of squandering so much on the Yard Seam things might have been so much better. It was of little use denying that matters stood very dangerously in this district. What could he do? When the pit was closed would it be best to wall the place in at some point along the main intake, or ought he to leave it.

John Tate knew the value of a good flow of air. He knew the acute danger of an explosion. He would be glad when the last man got to bank.

What is an explosion?

An explosion of gas in a mine is nothing great in itself. Gas accumulates in all kinds of mines. It is present in ironstone. It is easily disposed of in an ironstone mine. It is easily thrust into the main stream of ventilation, or it is ignited. On a Sunday morning Acts of Parliament are defied and bad accumulations of gas are ignited, and that is the end of the matter with the possible exception of a bit of displaced timber and a fall of shale. It would be the same in a coal mine, only there is something present to feed the fire engendered by the ignition of gas. Coal-dust feeds the fire. The initial explosion of gas is fed and increased into a colossal tornado by forcing the infinite particles of coal-dust into the atmosphere and making it into an inflammable gas. The fire thus rages along the gallery *against* the incoming flow of pure air and rears itself into something so powerful, so gigantic, so destructive, so awe-inspiring that it cannot be imagined. An explosion roaring along the enclosed passage at such an enormous velocity, tearing up timber, destroying track, and engendering such a terrible heat as to ignite timber and coal is something which passes man's powers of imagination.

Such things happen.

They happen seldom.

Scientists have devised schemes for preventing the holocaust of a mine explosion. A certain stone-dust if widely scattered over the main intakes nullifies the power of the coal-dust to rise and take part in a passing explosion. It is costly to obtain, and it adds to the cost of coal-production by virtue of the fact that men must spread it about the engine planes. In these days a wage of fifty shillings a week must be taken into consideration when profits are menaced.

So John Tate sat with his problem.

He smiled sardonically into the darkness about him. He had known that this gas had existed for some time. but he was safe. They could put nothing on John Tate. The Mines Act provided for local mines' inspectors — men appointed by the miners themselves to inspect the pit and to report any delinquencies. He'd sack the first obstreperous swine who reported gas in this district. The report sheets stated explicitly that no gas existed. It did not. It *dare* not.

Harri Webb
The Senghennydd Disaster

More than four hundred
Died at Senghennydd
Killed in one morning
By dust, firedamp, fire.

In the underground districts,
Mafeking, Kimberley, Ladysmith,
The York and Lancaster shafts
Of Sir William Thomas Lewis,
Baron Merthyr of Senghennydd,

Who postponed his holiday
As soon as he heard.
It was eight in the morning
In a small cold place.

They were identified only
By the patches on their rags,
Known to the women
Who'd washed them so often,

As they stood by the pit-head,
And on the bleak hillsides,
Everybody wondered
At their lack of tears

Four hundred and forty graves
Wet in the rain
Of seventy years,
The tears came later.

Seventy years of tears.
The sword rings in the heads of mothers.

John L. Hughes
A Bit of Safe Compo

And underground that night you breathe a lot of dust same as the
rest of Beano's gang. Setting up this new seam facing for the

firemen who don't like cutting their hands. Hacking and picking all manual till the small hours. Burrowing in deep through them Maritime vaults. Shifting skiff after skiff of hardcore rubble. Making you think as how you have drawn your last load of bitumen in that district. Cursing the surveyors up in heaps for saying there will be coal after eight to eleven feet. And still hacking rock after fourteen.

With a lot of coughing and spitting for hours there in the dark-brown gloom slashed occasionally through the dust by lamp beams and voices desperate for air.

They sent us the wrong bloody way.

There's no coal in this end.

Surveyor must be pissed.

He got his instrument twisted.

Thinks he's bloody chocolate.

Send the bastard down here.

Put a pick up his hole.

Bloody office boys.

But finding coal after fifteen feet running Vee pattern. And a seam they can blow out continuous for six months or more. Crawling back on your bellies to the last deadface where the gang sit down coughing in chorus. And mister I can tell you for nothing as how around this town they call that noise the national anthem.

Remembering Tommy and the way he went off to meet his maker two years past with lungs full of pneumoconiosis and blind in both eyes and deaf in both ears and a tongue that couldn't talk on account of all them wires holding his jaw together. Being Rachel's old man who fell into the grinder sorting good coal out from bad. Fed on baby food through plastic tubes. Looking like a Martian dribbling glucose from his agonised gob. And only then in between long sucks of oxygen up his nose behind this ugly green face bag you finally use when you got over ninety per cent inside them dust-filled organs.

Hearing still Rachel telling him the last words she would ever speak to his muted mangled face. And him not knowing she was in the room till she gives him a kiss on one of them blind eyes.

Ben have come to sit with you.

He have come for company.

That's a full cylinder.

You are OK now Dad.

The doctor says not to worry.

I'll give you a wipe.

There's like a baby you are.

You are OK now Dad.

Mam have gone to Ponty.

Ben have come to sit with you.

Staring at the poor bastard who never heard nothing of what she said. Staring at him. Sightless milky white eyes fixed on the ceiling. Gasping out his last. Creamy grey-faced up and down horribly in rhythm with his chest. Thin as a starved herring. A fading shadow of the man he once was judging by his photo on the dresser standing smiling fat twenty years ago outside Coney Beach fairground arm in arm with his collier opos on a sunny day.

Hearing the hiss from the cylinder valve now and then as he regulates his breath. And the smell of camphor strong on his Daz-white pillow-case. And he have never seen you. And you have never bought him a pint down the club nor anywhere. Yet in the loneliness of the fag-end of his life his hand reaches out for yours like something have told him you are there.

And his thin fingers grope across yours and he feels your nails and he feels your palm and he grips your thumb. And through all them jaw rivets and silver wires his lips part into something of a smile. Giving you the willies for certain on account of you knowing he is definitely not smiling because he is happy but because his fingers have told him you work down the pit.

And mister I can tell you for nothing as how you don't ever forget that smile. Like he is showing you the end of your life. And how you had better look hard while you still got eyes.

Except you are not stupid like he been. And they will never catch you near no grinder after your coal-pulling strength have gone. And the bastards will never shelve you off to woman's work same as they done to poor Tommy. Look what happened to him once he lost his self-respect.

Telling them deaf ears as how they will never keep you there underground as long as him. Just long enough to pay off the house and throw away that rope from around your neck. Just long enough to work a bit of safe compo same as plenty you know. Not like him. Never like that.

Robert Morgan
Blood Donor

The searching was easy and memory ripens
On the grey earth picture of Rees
In his grimy vest soaked in blood.

Forty-eight years under tense rock
Had stripped him like a tree with roots
In slag and marked him with texture of strain
And accident. But it was slow legs
And dust-worn eyes that were to blame.

The iron rock-bar was still in his hands
Held like a spear of a fallen warrior.
The rocks had dyed his silver hair red
And the heavy bar was warm and worn.
Blind flies swarmed in the blood-sweat
Air and the tough men with bruised
Senses were gentle, using distorted
Hands like women arranging flowers.

On the way out through roads of rocky
Silence you could sense images of confusion
In the slack chain of shadows. Muscles
Were nerve-tight and thoughts infested
With wrath and sharp edges of fear.
Towards the sun's lamp we moved, taking
Home the dark prisoner in his shroud of coats.

Myfyr Wyn
The Shape of the Rock

... My father died thirty-nine years ago. He met with an accident
at work in the Bryn-bach pit, Tredegar, and I was at his side. To
attempt to give a picture of my emotions at the time would be a
greater task than I would wish to undertake. His workmate
attempted to get me to fetch help while he tried to free him from
under the rock, but I could no more move from the place than if
I had been paralysed, and he had to turn very sharp with me before
he got me to shift. I went at last, and I had a considerable distance
to go before reaching the nearest workmates. Oh dreadful time!
It lives in my memory now, and as perfectly as if it had happened
yesterday, though at the time I was only ten years old. *The shape
of the rock* that fell upon him is struck upon my memory so deeply
that time cannot wipe it out. The place where the accident
happened and all connected with it are by me still, and my mind
often and often traverses the old roadways to 'pen y cwymp' in

the Garw seam, in Bryn-bach pit, where my father had the blow which caused his death within a little time, and which widowed my mother and left four of us young and fatherless. I can do no less than mention, too, my living memory of his being carried home on his workmates' shoulders, and the sympathy shown by the old inhabitants for him and myself. Everyone in the district knew him, and the tears and sighs I saw and heard struck me to the heart. It must be remembered that the old people of that time thought more of such circumstances and took greater notice than in the present cold and unfeeling age. The frequency of accidents now in our pits, and everywhere, makes us more off-hand and heedless than was the case then. Today, many of our neighbours are carried home fatally injured from the pits, and we know nothing about it until we read their story in the newspapers.

Dannie Abse
Heritage

A heritage of a sort.
A heritage of comradeship and suffocation.

The bawling pit-hooter and the god's
explosive foray, vengeance, before retreating
to his throne of sulphur.

Now this black-robed god of fossils
and funerals,
petrifier of underground forests
and flowers,
emerges with his grim retinue
past a pony's skeleton, past human skulls,
into his half-propped up, empty, carbon colony.

Above, on the brutalised,
unstitched side of a Welsh mountain,
it has to be someone from somewhere else
who will sing solo

not of the marasmus of the Valleys,
the pit-wheels that do not turn,
the pump-house abandoned;

nor of how, after a half-mile fall,
regiments of miners' lamps
no longer, midge-like,
rise and slip and bob.

Only someone uncommitted,
someone from somewhere else,
panorama-high on a coal-tip,
may jubilantly laud
the re-entry of the exiled god
into his shadowless kingdom.

He, drunk with methane,
raising a man's femur like a sceptre;
she, his ravished queen,
admiring the blood-stained black roses
that could not thrive on the plains of Enna.

Anon.
The Hartley Calamity

The Hartley men are noble, and ye'll hear a tale of woe,
I'll tell the doom of the Hartley men, the year of sixty-two.
'Twas on a Thursday morning in the first month of the year,
When there befell the thing that well may rend the heart to
    hear.

Ere chantecleer with music rare awakes the old homestead,
The Hartley men are up and off to earn their daily bread.
On, on they toil, with heat they broil, and streams of sweat
    do glue
The stour into their skins till they are black as the coals they
    hew.

Now to and fro the putters go, the waggons to and fro,
And clang on clang of wheel and hoof ring in the mine below.
The din and strife of human life awakes in 'wall' and 'board',
When all at once a shock is felt which makes each heart-beat
    heard.

Each bosom thuds as each his duds then snatches and away,
And to the distant shaft he flees with all the speed he may.
Each, all they flee by two, by three, they seek the shaft to seek
An answer in each other's face to what they may not speak.

Are we entombed? they seem to ask, for the shaft is closed,
     and no
Escape have they to God's bright day from out the night below.
So stand in pain the Hartley men, and swiftly o'er them comes
The memory of home and all that binds us to our homes.

Despair at length renews their strength, and they the shaft
     must clear,
And soon the sound of mall and pick half drowns the voice of
     fear,
And hark, to the sound of mall below, do sounds above reply?
Hurrah, hurrah for the Hartley men, for now their rescue's nigh.

Their rescue nigh? The sounds of joy and hope have ceased,
     and ere
A breath is drawn, a rumble's heard that drives them to
     despair.
Together now behold them bow, their burdened souls unload
In cries that never rise in vain unto the living God.

Whilst yet they kneel, again they feel their strength renewed
     again,
For the swing and ring of the mall attest the might of Hartley
     men.
And hark to the blow of the mall below, do sounds above reply?
Hurrah, hurrah for the Hartley men, for now their rescue's nigh!

But now that light, erewhile so bright, no longer lights the
     scene,
A cloud of mist the light has kissed and shorn it of its sheen.
A cloud of mist the light has kissed, and see, along does crawl,
Till one by one the lights are smote, and darkness covers all.

'Oh father, till the shaft is cleared, close, close beside me keep.
My eyelids are together glued, and I — and I must sleep.'
'Sleep, darling, sleep and I will keep close by — heigh ho — '
     To keep
Himself awake the father strives, but soon he too must sleep.

And fathers and mothers and sisters and brothers, the lover,
    the new-made bride,
A vigil kept for those who slept from eve to morning tide.
But they slept, still sleep, in silence dread, two hundred old
    and young,
To awake when heaven and earth have sped and the last
    dread trumpet sung.

B.L. Coombes
Twenty Tons of Coal

It happened three days ago. Three days have gone — yet my inside
trembles now as it did when this thing occurred. Three days
during which I have scarcely touched food and two nights when
I have been afraid to close my eyes because of the memory that
darkness brings and the fear which forces me to open them swiftly
so that I shall be assured I am safe at home. Even in that home I
cannot be at ease because I know that they notice the twitching
of my features and the trembling of my hands

That was why I forced myself to go along the street the day
after the accident. I wanted to go on with life as it had been before,
and I needed the comfort and sympathy of friends. The first I
saw was a shopkeeper whom I had known as an intimate for years.
He was dressing the window so I went inside to watch him, as I
had done many times.

I expect my replies to his talk about poor sales and fine weather
were not satisfactory for he turned suddenly and looked at me
before he said:

'Mighty quiet, aren't you? Looking rough, too. What's the
matter, eh? Got a touch of flu?'

'No! I wish it was the flu,' I answered, 'I could get over that.
I've had my mate smashed — right by my elbow.'

'Good Lord!' He is astounded for an instant, then remembers.
'Oh, yes. I heard something about it, up at that Restcwm colliery,
wasn't it? That's the way it is, you know. Things are getting pretty
bad everywhere. The toll of the road f'instance — makes you
think, don't it?'

'The roads,' I answer slowly, 'yes, we all use the roads. Can't
you realize that this is something different? He was under tons of
rock, and everything was pitch dark. No chance to get away; no
way of seeing what was coming; no — oh, what's the use? If you've

never been there you'll never understand.'

'Don't think about it,' he suggests, 'you'll get over it in time. Best to forget about it.'

Forget! The fool — to think that I can ever forget. I know that I never shall and no man who has been through the same experience ever can.

I went back home; soon afterwards one of my friends called. When he saw me he exclaimed:

'Holy Moses! What the dickens has happened to you?'

He had been with me the evening before that accident but that night had written such a story of fright and fear on my face that he could hardly recognise me.

So I stayed indoors, hoping that time would ease my feelings but jumping with alarm at every sudden word or slam of the door, and dreading the coming of each evening when the darkness of night would remind me of that black tomb which had held my mate but allowed me to escape.

Then again, this morning, after I had heard the clock striking all through the night, I must have surrendered to my exhaustion and slept, for I did not remember anything clearly after four o'clock struck. At five o'clock someone hammered on our front door. In an instant I was wide awake; the bed shook with my trembling. That crash on the door was the roar of falling rock; the darkness of the room was the solid blackness of the mine; and the bedclothes were the stones that held me down. When the knocking was repeated I had discovered that I was safe in bed. In bed — and safe; how can I describe what I felt.

Then I pondered what that knocking could mean. It was obvious that another morning was almost dawning. Griff, that was my mate's name, used to knock me up if he saw no light with us when he was passing to work. Could it be that he was passing: that all else had been a nightmare that this sudden wakening had dispelled? No, I realized it had been no nightmare for I had helped to wash his body — what parts it had been possible to wash without them falling apart.

Then came another thought; could it be that he was still knocking although his body was crushed? I dreaded to look, yet I could not refuse that appeal. I stumbled across the room, lifted the window, then peered down into the darkened street. A workmate was there. He lived some distance away but was on his way to the pit. He had a message for me. He shouted it out so that there should be no doubt of my hearing:

'Clean forgot to tell you last night, so I did. They told me at the office as you was to be sure to be at the Hall before four o'clock today. The inquest, you know. Don't forget, will you?'

Will I — can I — ever forget? Yet so indifferent are we to the sufferings of others that this caller, who is old enough to know better, who is in the same industry and runs the same risks, and who may be in exactly the same position as I am some day, does not realise how he has terrified me by hammering at my door to give that needless message.

Forget it! Is that likely when a policeman called yesterday and, after looking in my face and away again, told me gently that I was asked to be at Restcwm before four o'clock tomorrow — he had to say tomorrow then, of course. Not more than an hour later the sergeant of police clattered up to our door and — very pompously — informed me that I was instructed to present myself at the Workman's Hall, Restcwm, not later than four o'clock on the afternoon of Friday the, etc., etc.

After the caller has gone to work I get back into bed. I have been careful to put the light on because it will be a while before there is sufficient daylight to defeat my dread of the lonely darkness.

Be there by four — so I must start from here about two o'clock. Restcwm is a considerable distance away and I have other things to attend to before the inquest. I have to draw the wages for last week's work and I shall have to take Griff's to his house as I have been doing for years. Next week I shall be short of the days I have lost since the accident. I wonder if they will pay us for a full shift on the day that he was killed. I have been at collieries where one sixteenth of a shift was cropped from the men who took an injured man home just before the completion of the working day. I think our firm will not be so drastic as that; they are more humane in many ways than most of the coal-owners.

I lie abed, and think. The inquest will be this afternoon and I shall be the only witness except the fireman. This is the first time I have been a witness or had any connection with legal things and the police. I dread it all. I shall have to tell what happened in pitch darkness about two miles inside the mountain. They will listen to me in the brightness of the daylight and in the safety of ordinary life; and they will think that they understand. They may put their questions in a way that is strange to my experience and so may muddle me.

I shall have to swear to tell the truth, and nothing but the truth. Nothing but the truth, it sounds so simple. I will try to recall what happened and whisper it to myself in such a way that I shall be question-proof when the time comes.

Griff was there before me that night, as usual, sitting near the lamp-room. He gave me the usual grin at our meeting, then when he had finished the last of that pipeful and had hidden his pipe very carefully under that old coal-tram near the boilers, he took a last look at the moon, then we stepped into the cage and were dropped down.

I remember him saying as he looked upwards before we got under the pit wheels:

'Nice night for a walk ain't it, or a ride through the country. Nice night for anything, like, except going down into this blasted hole.'

Griff is many years older than I am; I expect he is about fifty. We have worked together for many years. He is well built but quite inoffensive. He has a couple of drinks every Saturday night and chews a lot of tobacco at work because smoking is impossible. He is aware that things are rotten at our job and is convinced that someone could make them a great deal better if they wished to; but who should do it or how it should be done are problems too difficult for Griff to solve. Soon he is going to have one of his rare outings; he is one of a club that has been saving to see Wales play England at Rugby football.

We were two of the earliest at the manhole where the fireman tests our lamps and tells us what work we are to do that night, for we are repairers and our place of work is changed frequently. The fireman is impatient and curt, as always.

'Pile of muck down ready for you,' he snaps each word and his teeth clack through the quid of tobacco as he talks 'there's a fall near the face of the new Deep. Get it clear quick. 'Bout eight trams down now, and you'd best take the hatchet and measuring stick with you because I s'pose as it's squeezing now.'

'Eight trams,' Griff comments as we move away, 'I'll bet it's nearer ten if it's like his usual counting.'

We hurry along the roadway, crouch against the side whilst four horses pass us with their backs scraping against the low roof, then move on after them. As we near the coal workings the sides and roof are not so settled as they were back on the mains. We hear the creak of breaking timber or an occasional snap when the roof above us weakens. The heat increases and our feet disturb the thick flooring of dust.

Where the height is less than six feet timber is placed to hold the roof but where falls have brought greater height steel arches are placed in position. They are like curbed rails, nine feet high to the limit and about the same width. Where they are standing we

can walk upright but we must be wary to bend low enough when
we reach the roof that is not so high. We have been passing engine-
houses as we moved inwards. These are set about four hundred
yards apart and become smaller in size as we near the workings.
Finally we pass the last one where the driver is crouching under
the edge of the arching rock.

The new Deep is the last right-hand turn before we reach the
Straight Main. Our tools are handy to our work and we are glad
to strip off to our singlets for they are sticking to our backs. We
see at once that the official was too optimistic for the fall blocks
the roadway and it is difficult to climb to the top of the stones.

'Huh,' Griff is disgusted, 'more like twelve it is. Eight trams
indeed. I guessed as much.'

It is squeezing, indeed. Stones that have been walled on the side
are crumbling from the pressure and there sounds a continual
crack-crack as timber breaks or stones rip apart. As we stand by,
a thick post starts to split down the middle and the splitting goes
on while we watch, as if an invisible giant was tearing it in half.
Alongside us another post that is quite an foot in diameter snaps
in the middle and pieces of the bark fly into our faces. The posts
seem no better than matchsticks under the pressure and we feel as if
we were standing in a forest — so close together are the posts — and
that a solid sky was dropping slowly to crush everything under it.

'Let's stick a couple more posts up,' I suggest, 'because most of
these are busted up. Perhaps it'll settle a bit by then.' We drag
some posts along the roadway, measure the height, then cut the
extra off with a hatchet that must not be lifted very high or it will
touch the roof. When we carry the timber forward we listen after
every step, with our head on one side and our senses alert for the
least increase in that crackling. We have measured the posts so
that they should be six inches lower than the roof, then the lid can
go easily between, but when we have the timber in position we
notice that the top has dropped another inch. When we are
tightening them we are careful not to hold on top for fear that a
sudden ease in the pressure may tighten it suddenly and fasten
our hands there. Ten minutes after the posts are in position
turpentine is running from them — squeezed out by the weight.

The journey rider — this one is called Nat — comes along and
we help him to repair the broken signal wires. He knocks on them
with a file; there is a bang and rattle as the rope slackens and a
tram is lowered to us. It seems that the roof movement is easing
a little so I climb on top of the fall to sound the upper top. I have
to stretch to my limit to reach it although I am standing quite nine

feet above the roadway. The stones above echo hollowly when I tap them with the steel head of a mandrel so we are convinced that they have weakened and may fall at any minute. The awkward part is that we shall have to jump back up the slope and that the tram will be in the way to prevent us getting away quickly.

One of these trams holds about two tons and we had to break most of the stones, so we were busy to get the first tram filled in the first half hour. Nat signalled it to a parting higher up where a haulier was waiting with his horse to draw it along the Level Heading where the labourers would unload it into the 'gobs'. Whilst Nat was lowering another empty tram we noticed the small flame of an oil-lamp coming down the slope.

'Look out, you guys,' Nat warned us when he stopped 'here's the bombshell coming and he'll want to know why the heck we ain't turned the place inside out in five minutes, you bet he will.'

It is the fireman and he came with a rush, stumbling over a loose piece of coal and almost falling; whereupon Nat turns away, partly as an excuse for not putting out his arm to steady the official and partly to hide the grin that he has started in anticipation of seeing the fireman go sprawling. The fireman recovers, however, and he glares at Nat as if he had read his thoughts. His hurry has caused him to breathe gaspingly; drops of sweat are falling from the end of his nose and the chew of tobacco is being severely punished. He glares at the fall, then back at us as if he thinks we must have thrown more on top of it.

'There's one gone,' I tell him, 'and a good nine left still.'

'Huh!' he grunts, 'don't be long chucking this one in agen. There's colliers below and coal waiting.'

He rushed away to hurry the labourers. We were full again when he returned in twenty minutes' time.

'While the rider's taking these trams up,' he ordered us, 'you roll some of these stones and wall 'em on the sides. Put 'em anywhere out of the way of the rails.'

Griff went to have a drink after we had filled the fourth. The water gurgled down his throat as it would have down a drain.

'Blinkin' stuff's got warm already,' he complained, 'and it was like ice when I brought it into this hole.'

His face is streaked with grey lines where the perspiration has coursed through the thickness of dust; when he wrings the front of his singlet the moisture streams from it. The fireman visited us every few minutes and upset us with his impatience. Even when he did not hurry us with words we could sense that he felt we were taking too long. It was nearly three o'clock in the morning when

Nat arrived with the tram that would be sufficient to clear the roadway. It seems that the mountain always becomes uneasy about that hour and small stones had been flaking down like heavy raindrops. We peered out from under the edge of the hole and I said that these falling stones must be coming from the upper edge of the right side. I could see some stones there that had half fallen and become checked in their drop. I got the slender measuring stick — it was about nine feet long — and tried to reach those loose stones but when the stick was to its limit and my arms were outstretched I could not reach the upper top. I climbed upwards on some of the stones that had been walled near the side. When I had scrambled up to about eight feet high it was possible to tap the stones and they fell. It was warm down below but the heat was intense up in the hollow of the fall. The increase of temperature almost stopped my breathing; I noticed the warning smell that is like rotten apples. My head was so giddy that I could not climb down; I slid the last part.

'Phew!' I gasped. 'It's chock full up there. My head's proper spinning.'

'Full? What d'you mean?' the fireman demanded, although he knew.

'Full up of gas,' I replied, 'and there's enough in that hole to put us up to the sky.'

'What are you chirping about?' he snapped back at me, 'there's nothing to hurt up there.'

'Try it and see,' I suggested. 'I notice you haven't tested for any tonight.'

'Get on and clear that fall,' he said, 'there's nothing there.'

'Take your lamp up there,' I insisted, 'it's the only oil-lamp here. I know the smell of gas too well to be sucked in over it.'

Very reluctantly he began to climb but when he was nearly up he jerked his hand and the light was extinguished.

I had expected it, for that was better than showing there was gas present and he was wrong.

'Now just look what you've done,' he complained, 'I'll have to feel me way back to the re-lighter.'

I saved my breath because I knew further comment was useless. The official stayed sitting on the wall like a human crow and watched us while we went on with our filling. We were about half-full when I heard a sound like a stifled sob and the fireman slumped down, then rolled to the bottom quite near Nat. The rider jumped back as swiftly as a cat, then crouched under the shelter of a steel arch.

'Now, where the devil did that 'un drop from,' Nat demanded. Then he turned to look at what had fallen. 'Good Lord!' he added, 'it ain't a stone — it's him. Out to the blinkin' world, he is. So there was some up there all right.'

Our lights showed us that the fireman was breathing, although faintly.

'Let's carry him back to the airway,' I suggested, 'there's a current of fresh air there and he'll soon come round.'

'Too blasted soon, likely.' Nat was not sympathetic. 'The only time this bloke is sensible is when he's asleep. And why struggle to carry him when I got me rope as I can put round his neck and the engine as can drag him?'

After we had carried him to the airway we went on with our job. We had two pairs of steel arches to place in position and bolt together. We were anxious to erect them so that we could cover them with small timber in case any more stones fell. Nat agreed to sit near the official and shout to us if there was any undue delay in his recovering.

'Fan him with me cap.' Nat was angered when I suggested it. 'Why the hell should I waste me energy on him, hey? Let him snuff it if he wants to, I shan't cry.'

He seemed to be looking forward with delight to the time when the fireman would open his eyes and see the sketches that were chalked on the smooth sides of that airway. We had some skilful artists in that district and no one could mistake who was represented as waving that whip behind those three figures who were carrying shovels.

We had the one arch solid and were well on with the second before the fireman recovered enough to stumble up the road towards us. He did not praise us for our speed in erecting, I do not think he was very appreciative of anything just then. He said nothing as he passed but climbed on to one of the tram couplings. Nat warned him to hold tight in a manner that showed that the fireman could fall off under the trams if he liked, then they moved away up the road away from us.

As soon as we had covered the steel arches we went to have our meal. We moved to where the roof was stronger, covered our shoulders with our shirts and sat on a large stone close to one another. We leant back against the walled sides, partly to ease the ache in our backs, and partly to lessen the target if more stones should fall.

We were supposed to have twenty minutes for eating food but we had finished before that. Griff looked at his watch; it was ten minutes to four. I remember him stating the time and remarking that we had not been disturbed at our food for a wonder. Hardly had he said that when we saw a light coming towards us. We could tell by the bobbing of the lamp that the one who was carrying it was running.

Whoever is coming it must be a workman because he is carrying an electric lamp. We can hear him panting as he comes and his boots hit the wooden sleepers with a thud.

'Something have happened.' Griff speaks my own thoughts, 'somebody have been hurt bad or —' He does not finish and we wait, tensed, for the message.... The running man reaches us, pauses, then holds his lamp up to our faces. The shadow behind the lamp becomes more solid and I realize that it is Ted Lewis.

'Puff,' Ted blows his cheeks out, 'all out of wind I am. Been hurrying like old boots to get to you chaps.'

Already we are reassured because if someone was under a fall Ted would have shouted his message at once. After taking a deep breath he explains:

'Old bladder-buster sent me to fetch you chaps to clear a fall he did. Said to come at once and bring your shovels and a sledge.'

'Fall!' We are both annoyed. 'Making all this fuss about a blessed fall.'

'Aye, I know,' Ted insists, 'but it's on the main and in the way of a journey of coal. He's in a hell of a sweat about it, not 'arf he ain't. Told me to tell you to hurry up — to run along with your tools, he did.'

'Run! Huh!' Griff is disgusted. 'I s'pose as we'd best go, eh? Allus something, there is.' With our tools under our right arms and our lamps hanging on our belts we hurry after Ted. We are careful to keep our heads down to avoid hitting the low places. Near the top of the third Deep we must meet the fireman, who swings around and walks in front of us. Suddenly he shouts back at us:

'There's ten full trams of coal the other side of this blasted fall and they won't be out afore morning if you don't shape yourselves.'

He is wasting his breath, for his threats and hurryings have lost their effect on us. His forcing is as much part of our working lives as the stones that fall or the timber that will break. Our lives are now a succession of delayed coal and falling roof; besides we are hurrying all we can. The sweat is dropping from our eyebrows; I feel it running over the back of my hand where a stone has sliced the skin away, it smarts as if iodine was smeared over it. The official stops suddenly and gasps:

'Just on by there. Not more'n four trams down and all stones, so it won't take you long. Look lively and get it clear.'

I judge the fall and decide that it is nearer six trams than four. My lamp shows me enough light to see to the top of the hole and to detect the stones that hang, half fallen, around the sides. There is a whitish glint over the shiny smoothness of the upper top. We call that type of roof the Black Pan; it will drop without the least warning.

'What's that smooth up above sounding like?' I ask.

'Not bad,' the fireman answers.

'Have you sounded it?' I ask.

'Course I have,' he answered, and I knew he had not.

'I'm going to do it for myself,' I stated, 'because you can't be too sure.'

I climbed on top of the fall, then tapped the roof with the measuring-stick. Boom — boom — it sounded hollow, as would a tautened drum. I scrambled back down.

'That upper piece is just down,' I said, 'it's ready to fall. Best to put some timber under it?'

'It's right enough,' the fireman insisted, 'and by the time we messes about to get timber here the shift'll be gone and it'll be morning afore we gets that coal by.'

'It would make sure that no more fell to delay us,' I argued, 'and it would be safer then.'

'And if you was to slam in it would be clear quicker,' he snapped, 'it seems as you're bent on wasting time.'

'I'm not,' I replied, 'only I wants to be safe as I can. It's my body, remember, and a man don't want more than one clout from a stone falling from as high as that.'

'Get hold on the sledge, Griff,' he orders, 'and make a start. This chap have got a lot too much to say.'

Griff looks at the official, then at me. He is hesitant.

'Griff can do what he likes,' I said, 'but I'm not working under that top until it's put safer.'

'You'll do as I tells you or you know what you can do,' the fireman snarls, 'and that's pick up your tools and take 'em out.'

'I'll do that too,' I replied and threw my shovel on the side, 'and what about you, Griff? Are you staying?'

'Don't know what to do, mun,' he mumbles. 'P'raps it'll stay all right until we have cleared this fall. We've done it afore, heaps of times. Let's pitch in and clear away as soon as we can.'

I know that Griff has allowed the thoughts of his wife and family to overcome his judgement.

'Aye, that's the idea.' The fireman is suddenly friendly. 'Slam in at it. You won't be long and I'll stand up on the side and keep me eyes on the top. If anything starts to fall I'll shout and you can jump back.'

I know well that before the word of warning could have formed in his throat it would be too late. Griff looks at me in an appealing way. He will not start without me, but I do not want to feel that I am responsible for his losing his job. I decide to risk it with him but to listen and watch most carefully.

We start to work, breaking the big stones and rolling them back one on top of another until we have formed a rough wall that is about a yard from the rail. The roof is quiet for a while and so we work swiftly. The fireman keeps very quiet because he can see we are working to our limit so that we can escape from under that bad piece, and he knows that the quieter he keeps the better we can hear. He sits on the wall holding his lamp high and looking continually upward.

We had cleared about half of the fall and I had finished breaking a large stone when Griff asked me for the sledge-hammer. Our elbows touched when I handed it to him. As he hit with the sledge I lifted a stone on to my knees but it slid down and dropped a couple of feet from the rail. I moved a short pace after it, bent, then began to lift it again. When I was almost straightened up I felt air rushing past my face; something hit me a terrible blow on the back. I heard a sound that seemed to start as a sob but ended in a groan that was checked abruptly. The blow on the back hit me forward. I felt to be flung along the roadway whilst my face ploughed through the small coal on the floor of the heading. I am sure that fire flashed from my eyes, yet I felt at the same time to be ice-cold all over. My legs were dead weights hanging behind me. When I breathed I swallowed the small coal that was inside my mouth. My nose was blocked with dust, so were my eyes. I felt about with my hands before realising that my face was against the floor and pushing down my arms to lift myself. I whimpered with relief when I found I could use my legs and so my back was not broken.

I could feel something running down my back; obviously it must be blood. Above, below, and around me everything is black with not the slightest sign of light to relieve it. So, whatever has happened the lamps must be smashed and we can have no help from them.

I had just managed to get to my knees and start to collect my thoughts when I heard a scuffle a few yards away. Suddenly a new sound pierced the darkness. It was a sort of half-scream, half-

squeal. At first I could not realise what this terrible sound meant;
I had never before heard a grown man squeal with fear.

'Quick! Quick! Get me out!' It was the fireman screaming, and
he sounded to be quite near to me. It seemed that he had been
caught but was still alive. I did not hear the least sound from Griff.
I collected my strength and shouted, 'Griff-oh! Are you all right?'
I am far more concerned about my mate than the official. Griff
was near me when I was hit. He was much more in the open than
the fireman, who had chosen a part that was sheltered along the
stronger side. I had no reply from Griff, but the fireman heard my
call and I heard him sobbing with relief at knowing that I am alive
and near to him.

'Come here, quick,' he appeals. 'I'm held fast over here. Get me
out before more comes. Quick!'

I listen for some seconds, trying to puzzle where Griff was
standing. I have lost all sense of direction. Am I nearly on top of
my mate or will I press a stone still harder upon him if I move in
that direction? While I hesitate the fireman restarts his screaming.
Small stones drip around me continually, like the early drops of a
shower of solid rain. Probably these are the warning that bigger
stones are loosening but I cannot see what is above or which way
to crawl and escape have lifted one eyelid over the other and the
water from that eye has cleared away most of the dust. I can now
open both eyes, but I can see no more than I could when my eyes
were fast closed. I crawl towards the fireman, guided by his
screams. Soon I find myself checked by what feels to be a mass of
stone. I climb upwards, scramble over the top, then slide down.
I call Griff again, softly, caressingly, as if to coax him to answer,
whatever has happened, but no reply comes.

I press my shoulder against the solid side of the roadway so that
it shall guide me, then I crawl forward, very slowly. The fireman
knows I am nearing him and directs my movements — continually
imploring me to hurry. Suddenly I touch something that is softer
and warmer than stone. I run my hand along and know it is a
human leg. My every nerve seems to grate when I decide it must
be Griff's and that he is dead.

'That's the leg.' The fireman's scream relieves me. 'There's a
stone on it as is holding me down. Lift it, quick.'

I feel for the stone and set myself to endure the pain of lifting.
I might as well have attempted to move the mountain, for three
attempts fail to shake the stone. The fireman is speaking near my
ear; he is frantic; begs me to hurry; screams at me as would an
hysterical woman. I feel about and find a stone that I can move

so I push it tightly under the one that holds the leg. This fresh stone will ease some of the weight and will stop any more pressure coming on his foot.

I have realized that I cannot do more until I have help. I must crawl and get others. I tell the fireman so, but he begs me not to go. I know there is no other way, so I turn around and feel my way over the stones. My fingers touch the cold iron of a tram-rail, but as there is no sign of a tram on that side I am assured that this is the right way. I crawl alongside that rail, running my fingers on it for a guide.

'Don't you be long,' he screams after me, 'for God's sake don't be long.'

Above, in the darkness, I hear a sound that resembles the ripping of cloth. It is this noise that stones make when they are being crushed and broken by the weight that is moving above them. I must hurry, so that the fireman may be saved and to see if there is any hope for Griff.

I drag myself along a few yards, rest some seconds to ease the pain, then drag along again. I must have crawled more than two hundred yards before I saw a light in the distance. I could not shout, so I had to crawl close to the repairer who was at work there. He was some seconds before he understood the message that I was croaking, then when he did he became so flustered that he wasted some time hurrying back the way I had come before he realized it was useless going by himself. I lay in the darkness while he ran back to call the help that came very quickly. Soon the roadway was brightened by the lamps of scores of men, who hurried along and took me with them, and this time we had plenty of light to see what had happened.

I could see that at least another twenty tons had fallen and the hole under which we had been at work was now huger than ever. The place was all alive again, creaking above and around us. Posts back in the gob were cracking — cracking — as if someone was firing a pistol at irregular intervals.

The fireman was as I had left him. He had his back against the side. His right foot was free but the left one was held tightly under a large stone. We could see no sign of Griff. They lifted a rail from the roadway, then used it as a lever to ease the stone off the fireman so that he could be taken back from the danger. He was only slightly hurt because the weight had only been sufficient to hold him and the main body of the stone was resting on others. He would not sit down but wandered amongst the men continually telling them of his own fright and moaning, 'Who would have expected this?' They lost patience at last and someone told him to sit down and not delay the work.

Above the men who strained to clear the fall, huge stones several tons in weight had started to fall, then had pressed against other stones that were moving and each had checked the downward movement of the other. They had locked each other in that position and now remained balancing — partly fallen — but the slightest jar or movement of the upper top would send them crashing down to finish their drop on the gang of men underneath. There was a continual rolling above us like thunder that is distant. Little stones flaked from the larger ones and dropped on the backs of the men as they worked below. Each time a stone dropped all the men leapt back, for a smaller stone is often the warning from a bigger one that is coming behind.

Several of the men stood erect, with their lamps held high and their eyes scanning the moving stones up above. They kept their mouths open, so that the warning shout should issue with no check. The others, busy amongst the fall, tumbling and lifting whilst they searched under the stones, did not hesitate when a warning came — they sprang backwards at once and made sure that no man stood directly behind the other to impede that swift spring.

Men can lift great weights when fear forces their strength. These stood six in a row, then tumbled big stones away until only the largest one in the centre was left. This one needed leverage, so a man knelt alongside to place the end of two rails in position; they had to be careful not to put the end on a man's body. Several men put their shoulders under the rails then they prised upwards. As the stone was slowly lifted they blocked it up by packing with smaller stones, then started to lift again. When the stone was two feet off the ground they paused; surely it was high enough. There was something to be done now that each man dreaded; then, as if their minds had worked together, two men knelt down and reached underneath. Very carefully they drew out what had been Griff.

We retreated with our burden and left the sides to do what crushing, and the roof to do what falling, they wished. The pain of my back had been severe all the while, when the excitement slackened I felt sick and could not stand alone. I leaned against one of my mates for support and he placed his arm around me gently, as if I was a woman.

We all know the verdict well enough, but refuse to admit it. Griff seems to be no more than half his usual size. Some one takes his watch from the waistcoat hanging on the side. They hold the shining back against what they believe is his mouth. Thirty yards away another stone crashes down on top of the others and the broken pieces fly past us whilst dust clouds the air. The seconds tick out loudly through that underground chamber whilst forty men watch another holding a watch; when he turns around someone lifts a lamp near so that they can see. The shining back is not dimmed. We had all known, yet somehow we had dared to hope.

As we are going outwards I notice that the fireman tries to isolate me; he wants to talk. I avoid him and keep in the group. Some distance along I hear a queer sound and look back to see that he has collapsed. His legs have given under him and he cannot stand. He is paralysed with fright. Two of the men place their hands under him and they carry him along behind the stretcher. They have to lean inwards to avoid the sides and bend their heads down because of the top. The fireman senses the hatred that is in all our minds and he sobs continually but no one asks him if he is in pain.

When we reach the main roadway the journey of empty trams is waiting. We place the loaded stretcher across one tram and four men sit alongside it. The fireman is lifted into another tram and the rest of us scramble on.

Suddenly the fireman tries to reassert himself.

'All of you going out,' he complains, 'didn't ought to go, not all of you. That fall have got to be cleared so's to get the coal back first thing.'

It was as if he had not spoken. The rider knocked on the signal wires. We start to move outwards slowly, for the engineer has been warned that it is not coal he is drawing this time. The fireman starts his mumbling again and we realize that he will tell the manager that the men refused to listen to him. Already he has started to cover his tracks.

Outside, it is dark and raining. The lamps in the pit-mouth are smeared where the water has trickled through the dust on the globes. There is a paste of oily mud and wet small coal that squelches under our feet. The official limps away to the office. We notice, and comment on the fact that he walks quickly and with hardly any difficulty. He gets inside the office and we hear him fastening the door before he switches the light on. He intends to be alone when making his report. We hand our lamps in, telling the lamp-men to note the damaged ones and we answer their inquiry as to 'Who is it this time?' They return our checks but put Griff's in a small tin box. A smear of light is brightening the sky but it is raining very heavily when we start on that half-hour's journey to his home. We feel our clothes getting wet on our bodies and the blankets on the stretcher are soaking. Water rushes down the house-pipes and it bubbles and glistens in the light of the few street-lamps.

All the houses near have their downstairs lights on, for news of disaster spreads quickly; besides, it is time for the next shift to prepare. The handles of the stretcher scrape the wall when we take the sharp turn of the kitchen door. This is the only downstairs room they have, so we prepare to wash him there. Neighbours have been busy, as they always are in this sort of happening. A large fire is burning, the tub is in, water is steaming on the hob and his clean pants and shirt are on the guard as if he was coming home from an ordinary shift.

I see no sign of Griff's wife. I remember her as small and quiet; a woman who stayed in her own home and was all her time tending to Griff and their five children. I do hear a sound of sobbing from upstairs and conclude that they have made her stop there, very wisely. Sometimes I hear the voices of the children too, but they are soft and subdued, as if they had only partly wakened and had not yet realised the disaster this dawn had brought to them.

I think that is all. I have re-lived that night many, yes a hundred times since it happened, and each time I have felt that I hated the fireman more. Had that stone hit my back a little harder I would have been compelled to spend the rest of my days in bed with a broken back — and would have to exist on twenty-six shillings a week as compensation. Had I been a yard further back I would probably be in similar state to Griff — then I would have been worth eighteen pounds, bare funeral expenses, as I would have been counted as having no dependents.

If I appear stupid at the inquiry, as a workman is expected to be, then I will answer the set questions as I am supposed to answer them and 'the usual verdict will be returned'.

Griff was my mate, and nothing I can do will bring him back to life again, but his wife and family are left. He would have wished that I do the best thing possible for them. If I remain quiet, they may be paid about four hundred pounds as compensation — which is the highest estimate of the value of a husband and father, if he is a miner. They will think that one of the usual accidents robbed them of the father, but if they are told he should not have died, it will surely increase their suffering.

If I speak what is true, the insurance company will claim that they are absolved from liability because we should not have worked there. Had we refused we should probably have lost our jobs. The insurance solicitor will be present — ever watching his chance — and will seize on the least flaw in the evidence.

So this afternoon I shall go to the office and draw two pay envelopes that should contain about two pounds sixteen each. One is mine, the other I will take to his house. There five silent children will be waiting whilst their dazed mother is being prepared to go to the Hall and testify that the crushed thing lying in the kitchen was her husband and that he was in good health when she saw him leave the house.

If the verdict is anything except 'Accidental Death' that pay packet may hold the last money she will have — unless it is the pension and parish relief.

Later, tonight, I shall have to face another fear, I shall have to go again down that hole and re-start work, but at four o'clock I will be at the inquest, shall kiss the Bible and speak 'The whole truth and nothing but the truth' — perhaps. Would you?

Duncan Bush
Pneumoconiosis

This is The Dust,

black diamond dust.

I had thirty years in it, boy,
a laughing red mouth
coming up to spit
smuts black into a handkerchief.

But it's had forty years
in me now: so fine
you could inhale it through a gag.
I'll die with this now.
It's in me
like my blued scars.
But I try not to think about it.

I take things pretty easy
these days, one step at a time:
especially the stairs.
I try not to think about it.

I saw my own brother: rising, dying
in panic, gasping
worse than a hooked
carp drowning in air.
Every breath was his last
till the last.

I try not to think about it.

Know me by my slow step,
the occasional little cough, involuntary
and delicate as a consumptive's,

and my lung full of budgerigars.

Richard Llewellyn
Then I Found Him

My mother looked at me and tried to smile, but her face was slack
with weakness, and her mouth kept pulling in jerks that were ugly.

'I wonder what has come to your Dada?' she said, and her voice
was like her mouth.

'I am going down now to see,' I said, and got up. 'Is Bronwen
out?'

'She is down at Iestyn's pit,' my mother said. 'She went with
Olwen, this afternoon, to take dry clothes for him.'

'But he was coming up through our shaft,' I said.

'He was too long underground,' she said, 'so they went to
Iestyn's pit in case he came back up there. The crowd was too big
down the bottom, here.'

'They should never have left the house,' I said.

'It was something for them to do,' she said, and then she was
crying, but not with tears.'

'Mama,' I said, 'no more thinking like this, is it? You are in
darkness and frightened. Come you, now. A light, and a cup of
tea, quick.'

'Leave me,' she said, and I never heard her sterner. 'Go to your
father.'

'Yes, Mama,' I said, and kissed her, and went from the house,
and ran down the Hill to the Three Bells.

Dai was with the boys, all in their working clothes and Dai's cleaner than any, sharp with creases from the cupboard shelves, and tight for him. They all had a glass and Dai gave me one that was three fingers deep with brandy.

'Come you, Huw,' he said, 'a health. To two good ones underground. Drink with love.'

We drank, and Dai seemed to have drunk only tea, but I was still coughing when we were down among the crowd at the pit-top, with Dai holding my arm, and fisting with his right, and the rest of us using picks and shovels to have a clear way. Over to the cage we went, with police making a baton charge from the boiler house to keep the crowd from us.

'Have anybody come up?' Dai asked the sergeant.

'No,' the sergeant said, 'and the water-gauge is still rising in there.'

'Right,' said Dai, and holding on to me until we were in the cage, 'there is good to be in my clothes, and ready for work again. Not a button to meet on my trews, see, and string to keep me tidy round the middle. I have got a belly like a sow through sitting to swill in that old bar.'

The cage swung gently, not quite on the bottom level, for the water was up to the waist, and we stopped it where it would rest dry, and jumped, one by one, into a black stillness of quiet ice, walking through to the pumps as though chained at the ankles. One of the pumps was damaged, but the other looked to be sound, and we started to work on them till the engineers gave a signal up to the surface.

Good, it was, to hear the voice of them, and to know that the waters were beaten.

'They had a good try, whoever it was,' the engineer said: 'No time to finish, thank God in His Goodness. There must have been more than a couple.'

'Cyfartha must have caught somebody at it,' Dai said. 'You will find the rat in the water, here. But where is Cyfartha?'

'I wonder did he chase the others?' Gomer said.

'No surprise to me,' Dai said. 'Let us find him. You have got the eyes. Come, you.'

So into the main we went, with candles high and splashing at the rats, with water to the chest in places and to the knees later. Then we came to the trouble.

The roof had fallen. Props had been weakened, and the pressure of water had torn away cogs as though made of paper.

'O, God,' Dai said, and feeling the rock with his hands, 'is he under this?'

'Is my father?' I said, and seeing my mother plain beside me.

'Come on,' Dai said, 'into it.'

Into it, yes, into it.

With fright chewing holes in me, and my mouth dry, and trembling, I went at it with the pick, and Dai doing the work of three beside me.

We had to smash through that dead weight of stone and clay, and carry it rock by rock and spadeful by spadeful out of our way, knowing that somewhere inside it my father or Cyfartha might be lying hurt, or dying, or dead.

As we worked we prayed, and between the prayers we cursed the heavy, dead, stupid hardness of the stone, and the thick, lifeless clay, and then prayed each time we strained to lever a bit of rock that some sign would be given to us that we were near.

But we had to work carefully for the roof was soft and with low rumbles to warn us that more would come down on top of us if we put a pick too far, or a shovel too high.

When we tired, others took our places, and when they were dropping, three more, until our turn came again, but all the time we were taking rocks away, or piling clay and muck. To the knees in water, and bent, for there was only four feet of head room, and knowing we must work fast, but held back because of the danger of a fall to make our work a waste of time.

The candles began to go, and a man went back for more, and something to drink, for we were dying down there, so hot it was, so filled with dust, and a scum of dust thick on the water, and mud to the calves and water rising fast as we worked downwards.

Dai was thick with mud, and throwing rocks from him as fast as he could have his hands on them, with a curse for each one, and his mouth in a wide line of hate, and his eyes mad through shining black muck, using the pick now, and nobody able to go near because of its bite, and throwing it down to pull out more rock, and toss it behind him, careless where it went as long as it was not in front.

Hour after hour we were down there, and with every yard, air getting colder and stiffening us, and the water rising to freeze us about the waist, until life was only a dig, and a pull and a carry and drop, and a crawl back, and a dig and a pull and a carry and drop, and a crawl back again.

And muscle screaming please to rest, if only to straighten shouting back, or stretch the torn palms of the hands.

But Dai Bando was up in front there, burrowing without a stop, working in darkness, feeling for rock with his hands, no sound only the sobs of his breath, and in his crouched back, a mightiness of threat to any who stayed even to hitch his trews.

Then Dai shouted, a high whisper of a shout, that sent ants crawling up my head.

'Cyfartha,' he was shouting, 'here is his coat, see.'

'Up in a stall road,' I said, for we had worked close to the wall and the coat was in the hole going up to the right.

'Clear the main, or the stall road?' Dai said.

Everybody stopped work.

If we went on up the main, we might be leaving Cyfartha and perhaps my father up in the stall road.

If we worked up the stall road they might be dying in the main in front of us.

I believe God the Father knows how you feel at such a time and sends a sign.

We had a sign, then.

We heard Cyfartha's pick hitting a signal on a rock.

Up in the stall road.

If we had gone on working, we should never have heard.

And Dai, who had never been in Chapel to pray since a boy, hit his hands together, and fell on his knees in the muck, crying like a woman.

'O God,' he said, 'with thanks I am, for this gift to me. Cyfartha is the blood of my heart. Have my eyes and my arms. I am thankful. In Jesus Christ. Amen.'

'Amen,' said we all.

'Give me the bloody pick,' Dai said, with new life. 'Stand away now.'

And the pick swung and struck as though he had just started.

'Mind the roof, Dai,' Gomer said, in fear, for the pick was driving deep, and the stone above us was growling.

'To hell with the roof,' Dai said, like an animal, 'God is with us, and bloody near time too.'

Behind us we heard men coming, and saw lanterns, with the manager and more of the men behind him.

'Right,' he said, 'you men can go to the surface. I am proud of you.'

But Dai went on picking and pulling, and none of us stopped.

'Come on,' he said, with sharpness, 'these men are fresh.'

'I will crush him in pieces,' Dai screamed, up in the narrow tunnel, 'I will have Cyfartha from here. Tell him to go to bloody hell, with him.'

And the manager knew, and the rock came back, and back, and Dai went up, and up, lying full length now, and a man behind, full length, and behind him, another, full length, passing rock and muck behind, one to another, with the roof touching our backs, and our bellies in blood from stones and black heat that was pain to breath, about us.

And Dai screamed again, a sound of terror, and of triumph, dulled by the tunnel and the heat and footage.

'Cyfartha,' he was screaming, 'Cyfartha. Back out.'

'Back out,' Gomer said, in front of me and his bootsoles came close to my face to bruise.

'Back out,' I said, to Willie, behind me, and I slid back, taking my rock with me.

'Back out,' Willie said, to his hind man.

Out of the heading we crawled, and Gomer coming to fall in a faint in the water, and then Dai.

If the Devil rises from the Pit as Dai came from the tunnel, a few of us are booked to die a second death with fear.

Black, and naked, and with lumps of mud stuck to his head and shoulders, and all of him shaking with strength that has gone weak, he shone wet in the lantern lights, and his eyes framed with pink, sightless with tears, and his mouth wide to the roof to breathe.

And in his arms, Cyfartha, back, too, and still.

'Is my father up there?' I asked him.

'Up there,' Cyfartha said, but only just. 'I was after him.'

'I will take Cyfartha to the top,' Dai said, 'and back, then, for your father.'

'I am going in,' I said.

'I love you as a son,' Dai said. 'Go on.'

So up I went, and as far as Dai had gone, in a little chamber of rock, and more rock piled in from again.

'Dada,' I shouted, 'are you near me?'

I hit my pick on stone and listened.

Only the growling up above and voices from behind in the tunnel.

So on I went again, pick and pull, pick and pull and wasting more time getting the rock back, and scooping mud, and trying to shovel.

And then I found him.

Up against the coal face, he was in a clearance that the stone had not quite filled.

I put my candle on a rock and crawled to him, and he saw me, and smiled.

He was lying down, with his head on a pillow of rock, on a bed of rock, with sheets and bedclothes of rock to cover him to the neck, and I saw that if I moved only one bit, the roof would fall in.

He saw it, too, and his head shook, gently, and his eyes closed.

He knew there were others in the tunnel.

I crawled beside him, and pulled away the stone from under his head, and rested him in my lap.

'Willie,' I said, 'tell them to send props, quick.'

I heard them passing the message down, and Willie trying to pull away enough rock to come in beside me.

'Mind, Willie,' I said, 'the roof will fall.'

'Have you found him?' Willie asked me; and scraping through the dust.

'Yes,' I said, and no heart to say more.

My father moved his head, and I looked down at him, sideways to me, and tried to think what I could do to ease him, only for him to have a breath.

But the Earth bore down in mightiness, and above the Earth I thought of houses sitting quiet under the sun, and men roaming the streets to lose voice, breath, and blood, and children dancing in play, and women cleaning house, and good smells in our kitchen, all of them adding more to my father's counterpane. There is patience in the Earth to allow us to go into her, and dig, and hurt with tunnels and shafts, and if we put back the flesh we have torn from her and so make good what we have weakened, she is content to let us bleed her. But when we take, and leave her weak where we have taken, she has a soreness, and an anger that we should be so cruel to her and so thoughtless of her comfort. So she waits for us, and finding us, bears down, and bearing down, makes us part of her, flesh of her flesh, with our clay in place of the clay we thoughtlessly have shovelled away.

I looked Above for help, and prayed for one sweet breath for him, but I knew as I prayed that I asked too much, for how were all those tons to be moved in a moment, and if they were, what more hurt might be done to others.

Afraid I was, to put my hands with tenderness upon his face, for my touch, though with the love of my heart, might be an extra hurt, another weighing, for they were with dirt and cuts, and ugly with work that was senseless, not good to put before his eyes, for they were the hands of the Earth that held him.

His eyes were swelling from his head with pain and his mouth was wide, closing only a little as with weakness, and then opening wide again, and his tongue standing forth as a stump, moveless, dry, thick with dust.

And as the blood ran from his mouth and nose, and redness ran from his eyes, I saw the shining smile in them, that came from a brightness inside him, and I was filled with bitter pride that he was was my father, fighting still, and unafraid.

His head trembled, and pressed against me as he made straight the trunk of his spine and called upon his Fathers, and my lap was filling with his blood, and I saw the rocks above him moving, moving, but only a little. And then they settled back, and he was still, but his eyes were yet beacons, burning upon the mountain-top of his Spirit.

I shut my eyes and thought of him at my side, my hand in his, trying to match his stride as I walked with him up the mountain above us, and I saw the splashings of water on his muscled whiteness as he stood in the bath, and the lamplight on his hands over the seat of the chair as he knelt in prayer at Chapel.

Air rushed from his throat and blew dust from his tongue, and I heard his voice, and in that strange noise I could hear, and from far away the Voice of the Men of the Valley singing a plain amen.

So I closed his eyes and shut his jaw, and held him tight to me, and his bristles were sharp in my cuts, and I was heavy with love for him, as he had been, and with sadness to know him gone.

'We can move the rocks now, Willie,' I said.

'O Christ, Huw,' Willie said, 'is he out, then?'

'Yes,' I said, and feeling warmth passing from between my hands, 'my Dada is dead.'

'Hard luck, Huw, my little one,' Willie said, and coming to cry. 'Hard old bloody luck, indeed. Good little man, he was.'

My mother sat in the rocking-chair with her hands bound in her apron, and looked through the open doorway up at the mountain-top.

'God could have had him a hundred ways,' she said, and tears burning white in her eyes, 'but He had to have him like that. A beetle under the foot.'

'He went easy, Mama,' I said.

'Yes,' she said, and laughed without a smile. ' I saw him. Easy, indeed. Beautiful, he was, and ready to come before the Glory. Did you see his little hands? If I set foot in Chapel again, it will be in my box, and knowing nothing of it. O, Gwil, Gwil, there is empty I am without you, my little one. Sweet love of my heart, there is empty.'

Well.

Meic Stephens
Hooters

Night after night from my small bed
I heard the hooters blowing up and down the cwm:
Lewis Merthyr, Albion, Nantgarw, Ty-draw —
these were the familiar banshees of my boyhood.

For each shift they hooted, not a night
without the high moan that kept me from sleep;
often, as my father beyond the thin wall
rumbled like the turbines he drove at work, I

stood for hours by the box-room window,
listening. The dogs of Annwn barked for me then,
Trystan called without hope to Esyllt
across the black waters. Ai, it was their wail

I heard that night a Heinkel flew up the Taff
and its last bomb fell on our village;
we huddled under the *cwtsh*, making
beasts against the candle's light until the sky

was clear once more, and the hooters
sounded. I remember too how their special din
brought ambulances to the pit yard,
the masked men coming up the shaft with corpses

gutted by fire; then, as the big cars
moved down the blinded row on the way to Glyntâf,
all the hooters for twenty miles about
began to swell, a great hymn grieving the heart.

Years ago that was. I had forgotten
the hooters: my disasters, these days, are less
spectacular. We live now in this city:
our house is large, detached and behind fences.

I sleep easily, but waking tonight
found the same desolate clangour in my ears
that from an old and sunken level
used to chill me as a boy — the inevitable hooter

that paralyses with its mute alarm.
How long I have been standing at this window,
a man in the grown dark, only my wife
knows, as I make for her white side, shivering.

Barry Hines
On 15's Coal Face

The pit baths were divided into two changing areas; the clean side and the dirty side. When the miners arrived for work, they undressed and left their street clothes in a locker in the clean side, then walked through the shower room to the bays of lockers at the dirty side, where they kept their pit clothes.

The other rippers who worked in 15's tailgate, were nearly ready when Syd and Harry reached their lockers in the dirty side. Syd unlocked his door and took out his pit shirt. By the time he had put it on and rolled up the sleeves, his hands were filthy. He sat down on the metal bench which ran across the bottom of the lockers, and took a pair of green and white football socks out of his snap bag. The others did not notice them while they were balled up but as soon as he pulled on the first one, Albert said, 'Hey up, Syd, they're bobby-dazzlers aren't they? Where've you got them from?'

Syd neatly folded the white stocking tops as if he was getting changed for a football match.

'I'm thinking of making a come-back. I'm going to get back into training.'

'Get back into bed more like.'

Syd stood up, and, dressed only in shirt and socks, started to run on the spot. Alan, who was strapping on his knee pads, turned his head to watch him.

'Well, they're crying out for players at Sheffield, Syd. Both teams.'

'I could still do it, lad, don't you worry about that.'

He tried to touch his toes, keeping his legs straight. Harry sat on the bench and pulled on his boots.

'Gi'o'er now, Syd. You're wearing me out just watching you.'

After three attempts, none of which took him within six inches of his feet, Syd stood up and gripped his wrist.

Ronnie said, 'What are you doing now then, Syd, taking your pulse?'

'Dynamic tension this.'

'What's that supposed to be?'

'You know, Charles Atlas. Him who sent Harry his money back.'

He was still doing exercises when Tony walked past the end of the locker bay and saw him. He stopped and waited for his father to come back to normal before speaking to him.

'Hey up, what you doing with them socks on? They're mine.'

Syd, flushed and breathing heavily, was glad of the interruption. He could stop now without losing face.

'You weren't using them. They've been stuck in that drawer ever since you left school. Somebody might as well get some wear out of them.'

'I know, but what if I want 'em though?'

Syd reached into his locker and took out his trousers.

'Want 'em? What you going to want 'em for?'

'You never know. I might want to start playing again some time.'

'Ar, I know. There's only one thing you're bothered about playing, and it's nowt to do with football.'

The others laughed. Tony didn't and neither did Syd. They faced each other down the bay for a few seconds, then, with a slight jerk of his head to show his father how disgusted he was with him for showing him up, Tony walked away.

Alan nodded after him. 'He was a good footballer at school you know, your Tony.'

Syd, who was still looking towards the end of the locker-bay where Tony had been standing, said nothing. Harry said it for him.

'He was a better cricketer though. He should have kept on with it. We've seen him take some wickets, haven't we, Syd? He's the fastest bowler I've ever seen as a lad.'

Syd took his boots and jacket from the locker and closed the door.

'He's too busy courting, that's his trouble. There's only one wicket he's bothered about now, and that's his middle 'un.'

While the others chatted and told jokes, Syd put on the rest of his pit clothes in silence, his good humour spoiled by the exchange with his son.

[...]

On 15's coal-face, the two fitters had almost finished replacing the cowl on the panzer motor. Frank Morris left Steve to tighten up the bolts, then crawled off the face to go and inspect the broken air pipe in the main gate.

He had only just started working on it, when Steve crawled to the end of the face and shouted to him to switch the power on, so he could give the motor a try.

Frank connected the power cables to the transformer and dropped the switch.

Syd was walking back up the tailgate with Alan to fetch another ring. He turned to face the terrible roar, saw a blue flash, was lifted by the burning blast of air.

Harry felt the ground shake, heard the rushing air, was knocked down by it.

Tony and Jimmy heard a bump, felt the air move and stopped work. The machine man switched off the shearer and everybody starting shouting up and down the face, asking what had happened. Then, without waiting to find out, they began to crawl rapidly between the double rows of supports to get off the face and out into the gates.

Harry used the conveyor to pull himself to his feet. He felt at his head and looked at his fingers, but it was too dark to see them because the light on the roof was out. He felt for his lamp dangling from the battery on his belt, and tried to switch it on to look for his helmet. It was already on, but the dust was impenetrable and it smothered the beam, so he thought the bulb must have broken when he fell. Coughing and choking, he felt his way to the tannoy and shouted into it that there had been an explosion. Nobody answered. He shouted along the roadway: no reply. So, blind and almost suffocating, he stumbled off down the roadway, to try to find his way to the pit bottom.

Jimmy tried the tannoy in 17's maingate, but there was no answer there either. The men knew then that something was wrong, and they had to get out. They picked up their snap bags and jackets and started up the gate.

When they reached the junction at the end of the gate, they could smell smoke, and the amount of dust in the air made them cough and spit. They hurried on in the direction of the pit bottom, fighting to contain their fear, wanting to run, but resisting in case it was nothing serious and they started an unnecessary panic. They met other men leaving other workings. They all wanted to know what had happened, but nobody knew.

The further they walked, the safer they felt. The air gradually cleared, and when they finally pushed their way through the last air doors, and walked the last hundred yards to the pit bottom, there was nothing to indicate, except for the large number of miners gathered there, that anything unusual had happened: the lights were just as bright, the walls as white, the airflow just as fresh and strong.

But above ground, Forbes had immediately ordered the emergency procedure into operation. He had instructed the pit to be cleared, and the Area mine rescue teams had been alerted and were already on their way.

Tony stood in the pit bottom and looked for his dad amongst the waiting men. He could not see him. Perhaps he had already gone up? Perhaps he had not arrived yet? Wherever he was, it was no good staying here. They would know more on the surface. So he crowded into the cage when his turn came, the onsetter closed the gates, and the cage rose out of sight, watched by the others still waiting there.

In the pit yard, groups of miners still in their pit muck stood about in the sunshine. Every time the cage reached the surface they stopped speculating and waited for the men to come down the wooden staircase from the pit bank to ask them the news. What they were told they already knew. There had been an explosion. Nobody knew of any casualties. That was all.

Tony stood with Jimmy and the other men from 17's coal-face. He still could not see his dad. Perhaps he had gone to the baths? He would go and look. No, he would stay and watch the steps. He would be coming up on a later draw; he had been working further inbye than Tony.

The wheels of the winding gear continued to turn, the cage continued to bring men to the surface; but Syd was never among them. One group included an injured man, who delayed the queue as he was helped down the narrow steps by two mates. They held him by the arms, and he kept feeling at his head then looking at his hand. When they reached the bottom of the steps, Tony recognized his uncle Harry and ran across the yard to meet him.

He did not say anything at first. He could not stop looking at the blood seeping from under Harry's hair, and moving in a thick dark stripe down his cheek.

'What's happened, Uncle Harry? Have you seen my dad?'

Harry turned towards the voice and struggled to place it. Tony had seen that bleary look before, in the eyes of drunkards. Harry shook his head. He seemed too tired to speak. His assistants hadn't seen Syd either, so Tony went back to Jimmy and the others, who were listening to a man who had just come up on the same draw as Harry.

'... I wondered what the bloody hell had happened. We were on the paddy, then there was this thump and this terrific rush of air and everything went black. You couldn't see a thing. I thought I was a goner I can tell you. I thought my lamp had gone out an' all till I held it in my hand. Then I realized it was the dust, it was that thick, it just blotted the beam out completely. I mean, you can't believe it can you?'

They shook their heads, but more in sympathy than agreement

for, being miners, they could believe happenings much more frightening than that.

Jimmy said, 'Does anybody know what's happened yet?'

'Somebody said it was on 15's. I don't know if it's right though.'

'15's! My dad's working in the tailgate on 15's.'

The man who had been riding on the paddy train did not know who Tony was, but the boy's reaction, plus the cautionary expressions of the other face workers told him that he had said something wrong. Jimmy could see that Tony was upset. He was either going to start crying or start fighting. Jimmy stepped in front of him in case there was any trouble.

'Look, it's no good jumping to conclusions. Nobody knows what's happened yet.... Anyway, there's a rescue team here look, they'll know for definite when they've been down....'

And the dramatic arrival of the yellow rescue van immediately postponed any further speculation. It drove into the pit yard as fast as a fire engine, and even before it had stopped, the back doors were thrown open and the rescue team jumped out with their breathing apparatus strapped to their backs, and their helmets and kneepads on, ready to go underground.

[...]

Underground, the Doncaster rescue team moved cautiously into 15's tailgate in search of the four rippers. They had tested for, and located gas and were advancing in single file wearing their breathing apparatus. The force of the explosion had blown the conveyor all over the roadway, and many of the rings and tin sheets between them had been blasted out of place. The rescuers had to climb over debris every yard of the way, and where the roof had collapsed, they negotiated the jagged heaps hurriedly, fearful of another fall from the unsupported rock above. The stretchers were awkward in these conditions. There was no straightforward carrying, and the continual lifting and passing over obstacles slowed them down. The oxygen packs made their backs ache, the nose-clips and mouth-pieces became uncomfortable, and the effort and tension of it all made the sweat drip from their faces and dribble down their bodies and legs.

They were finally stopped by a fall which completely blocked the roadway. They shone their lamps over it and two of them climbed the shaley slope to see if they could get over the top. But the fall reached right up to the roof, and they came down with no idea of how far back it went. It could be a few yards thick, or it might be solid right back to the ripping edge. They had no way of telling. All they could do now was go back and report it.

Because they were coupled up to their breathing apparatus they could not speak, and had to communicate by horns attached to their belts. The captain honked his five times, and the six men started on the tortuous obstacle course, back up the tailgate to the fresh air base.

[...]

While the Doncaster rescue team was struggling back up to 15's tailgate to the fresh air base, the Wakefield team was on its way to investigate the maingate. They had detected no gas yet, the roadway was undamaged, and as they were travelling down a steep incline, they were able to make good time. Then they heard what sounded like a waterfall ahead. They could not see it yet, but the increasing noise quietened them and made them apprehensive as they approached it.

[...]

At the bottom of the drift, they were halted by a pond. Water had burst through the tin sheets and was pouring from a jagged fissure in the roof. The Overman, who was guiding the rescue team, shone his lamp around the damage and across the water to the other side.

'The explosion'll have knocked the pump out of action. What do you want to do, go back up the return and round 12's to see if we can get that way?'

The captain of the rescue team focused his lamp on the bent tins protruding from the water.

'How long would that take?'

'About half an hour I should think. Probably a bit more.'

'We'd better try this way first then. Wait here till I see how deep it is.'

He handed the canaries' cage to one of the others, then walked into the water cautiously, unsure of his footing.

'Bloody hell; it's colder than the sea at Blackpool.'

'Watch that jellyfish, George. Don't let it get behind your kneepads.'

The captain's kneepads disappeared under the water. It reached his waist. He made a detour to avoid the downpour from the roof, then took off his breathing apparatus and held it clear of the water as it reached his chest.

'It looks as though we're going to need Sea Link, George.'

The captain was too short of breath from the encroachment of the cold water to answer; but it became no deeper and the submerged parts of him reappeared as he waded up the slope and out of the water at the other side.

'Right, lads, in you go. Take your equipment off and keep it well up out of the wet.'

The Overman shone his lamp across the water on to the captain standing there in his teeming overalls.

'Listen, I'm going back to see if I can get through on the blower. I'll try and get somebody down to have a look at the pump. See if we can shift this lot.'

He turned and walked away, back up the drift. The rescue team started to walk into the water, one after the other. One man pulled a face when the water went over his boots.

'Jesus Christ.'

The man behind him, who was still dry, shone his lamp on his mate's face, and was dismayed by the expression he saw there.

'I wish I was. I'd only get the bottoms of my feet wet then.'

When there was a line of them in the water, they relayed the stretchers and canaries to the other side. This jerky passage agitated the birds, and as the swaying cage was handed on, they jumped from perch to perch until they reached the captain and were placed on the ground.

[...]

Still soaking from the pond at the bottom of the drift, the Wakefield rescue team moved cautiously up 15's maingate towards the coal-face. The air was clean here, and the men were not using their breathing apparatus; but the leader still held up the bird cage every few yards to check the condition of the canaries inside.

They stopped and shone their lamps over an air door which had been blown down by the explosion.

'Christ, it must have been some blast to blow that off like that.'

They walked over the door and the team captain shone his lamp along the wall.

'It looks as though the flames have travelled this far. There's signs of coking here, look.'

He showed them the coal dust which had been burnt by the passing flame, and one of the men scraped some from the wall and crumbled it between his fingers.

As they turned to carry on, there was a shout from the darkness somewhere ahead.

'Hear that? There's somebody up there!'

They moved quicker now, but not recklessly, and they tested for gas as regularly as before they had heard the shout. They heard it again, much closer this time, and as their beams searched ahead, they lit twisted steel, rock, conveyor parts, then a man lying on his back in the side of the roadway. They ran the last few yards to reach him.

His helmet had been blown off and his clothes were scorched rags. One side of his face was raw flesh where the skin had burned away, and the boot on his right foot was sticking out at an unnatural angle. He raised his head and looked for the faces behind the lights.

'What time is it?'

[...]

The four remaining members of the Wakefield rescue team worked their way down 15's maingate, looking for Frank Morris and Steve Oates. The lead man shone his lamp on something on the floor and picked it up. The others gathered round him to look what it was.

'It's a page out of a book.' He turned it over and held his lamp closer. 'They look like race horse names to me.'

He dropped the page and they carried on. A few yards further on they found another one. This time the captain picked it up. Some of the horses' names were obscured by blood. He screwed it up and threw it down, and when they found a third page they left it.

The loose pages became more frequent, and led them to what looked like a snatch of material on a conveyor roller. The captain bent down to examine it, then recoiled as if he had reached into a hole and grasped something live.

'Bloody hell!'

'What's up? What is it?'

'It's somebody's arm.'

They left it lying there and, close by, they found a shattered body with bloody pages scattered round it. The sight of it was too much for one of the rescuers, a new member of the team who had never attended a disaster before, and he had to turn away and support himself against the wall.

While the captain searched the dead man's clothing for identification, the two healthy members of the team picked up the bird cage and moved the last few yards to the entrance of the coal-face. There was the sound of stone and metal being moved, then a shout, 'The other one's here!'

The man at the wall was violently sick. One of the men at the face called, only quieter this time, 'He's dead an' all!'

The captain stood up from the first body.

'There's no way of identifying him. I can't find his check or owt. He's blown to bloody pieces, there's only his clothes holding him together.'

He approached his distressed mate.

'Are you all right, Maurice?'

'I am now.'

'I know how you feel, lad. We've all gone through that at one time or another. Just hang on here a minute while I go and see what the others are doing.'

When he reached them, they were digging out Steve Oates's body with their bare hands. The captain helped them, and after they had freed it and laid it in the gate, he said, 'Poor bugger, he's nowt but a lad.'

They went back to the face and directed their beams into the low tunnel. They lit fallen rock and shattered machinery, but the fall was not solid like the one in the tailgate.

One of the men whistled softly at what he saw. He said, 'Bloody hell, we're going to have some pain on breaking through that lot. I hope there's plenty of men available.'

They crawled back into the maingate and stood up. The captain said, 'Right, let's get back and report then.'

They walked back to Maurice, who had recovered now, and he helped them to strap Frank Morris's body to the stretcher. Their other stretcher had been used for Ken Taylor, so they had to leave the apprentice's body at the coal-face for the next rescue team to carry out.

[...]

Tony's team was hard at work on the fall in the tailgate. The five men had formed a chain, and were passing rocks down the heap to be stacked along the tunnel walls at the bottom. Tony was on top of the heap. He was stripped to the waist now, and the mixture of sweat and dust gave his body an oiled and streaky appearance.

The Overman looked at his watch and walked up to the fall.

'Right, lads! Change over!'

The other members of the team stopped work and walked back up the gate, thankful for the rest: but Tony continued dragging at the rocks and sliding them down the slope as if he had not heard.

'Come on, Tony! Give it a rest now!'

Tony stopped work and knelt there panting.

'I'm all right, Reg. I'm just getting warmed up.'

'Listen, you'll be knackered if you carry on at this pace. It'll catch up with you in the end. You've got to look after yourself. What's the point in knocking yourself up in the first few hours, when we might be at it all night?'

Tony thought about it, then did as he was told and came down off the fall.

'Do you think we will be?'

The Overman studied the obstruction before them. The tons of rubble which had already been removed had made little difference to its size.

'There's no telling, lad. We'll just have to keep going, that's all.'

A fresh team started work, and Tony walked back to where his mates were resting. They were sitting on the floor with their backs against the wall and they were still breathing heavily from their stint at the fall. Tony took a bottle of water from his jacket pocket and had a long drink. As he screwed the top back on, he looked to see how much water he had left, and wondered how long he would have to make it last.

'What I don't understand is, what they're searching the main-gate for if the explosion's happened here.'

Jimmy opened his eyes and looked up at him.

'Just because there's a fall here, doesn't mean that the explosion's happened here, Tony. The blast gathers intensity, and you often get what they call a quiet zone where the explosion occurred. Surprisingly enough, there's not a lot of damage there, but it gathers force and your worst damage can be hundreds of yards away.... Anyway, they'll be trying to get down the maingate, to see if they can get round the back of this fall by crawling along the face.'

Tony put his bottle away, then sat down beside the machine man, who smoothed out a little space for him on the stony floor.

'At least there's no fire, that's one good thing. A mate of mine worked at Gressley when the pit got on fire that time. They'd to sandbag the whole district off to stop it spreading. It burned for a week ...'

He patted his trouser pockets to locate his snuff.

'... Once anybody gets caught in owt like that, they've no chance. Even if they don't get burned, they choke to death in no time.'

He took off the tin lid and offered Tony the first pinch. He refused and looked towards the fall.

'Do you think there'll be any air getting through in there?'

They watched the men shovelling and pulling at the rocks.

'You can't say. We'll just have to hope for the best and work like bloody hell until we break through.'

They all worked and waited through the night. Then, as night was ending, but before the first bird had begun to sing, the Wakefield rescue team crawled off the coalface and hurried back up the maingate to where the relief teams were resting.

[...]
The others heard them coming and watched their approaching lamps. One man took his watch out of its protective tobacco tin. 'Bloody hell, it's not time to change already is it? They've only been gone ten minutes.'
The Wakefield men were shouting the news before they arrived. 'We're through! We can get through! It's clear!'
Everybody jumped up and started talking and asking them questions when they reached base, but the superintendent in charge of the base quietened them down, and organized who should go back down the gate, and crawl along the face to try to reach the rippers trapped behind the fall in the tailgate.
Unaware of the breakthrough, the Milton rescuers worked on at the fall from the other side.

Leslie Norris
Elegy for David Beynon

David, we must have looked comic, sitting
there at next desks; your legs stretched
half-way down the classroom, while
my feet hung a free inch above

the floor. I remember, too, down
at The Gwynne's Field, at the side
of the little Taff, dancing with
laughing fury as you caught

effortlessly at the line-out, sliding
the ball over my head direct to
the outside-half. That was Cyril
Theophilus, who died in his quiet

so long ago that only I, perhaps,
remember he'd hold the ball one-handed
on his thin stomach as he turned
to run. Even there you were careful

to miss us with your scattering
knees as you bumped through
for yet another try. Buffeted
we were, but cheered too by our

unhurt presumption in believing
we could ever have pulled you down.
I think those children, those who died
under your arms in the crushed school,

would understand that I make this
your elegy. I know the face you had,
have walked with you enough mornings
under the fallen leaves. Theirs is

the great anonymous tragedy one word
will summarise. Aberfan, I write it
for them here, knowing we've paid to it
our shabby pence, and now it can be stored

with whatever names there are where
children end their briefest pilgrimage.
I cannot find the words for you, David. These
are too long, too many; and not enough.

Anon.
The Gresford Disaster

You've heard of the Gresford disaster,
And the terrible price that was paid.
Two hundred and forty-two colliers were lost,
And three men of a rescue brigade.

It occurred in the month of September,
At three in the morning, that pit
Was racked by a violent explosion
In the Dennis where gas lay so thick.

The gas in the Dennis deep section
Was packed there like snow in a drift,
And many a man had to leave the coal-face
Before he had worked out his shift.

A fortnight before the explosion,
To the shotfirer, Tomlinson cried:
'If you fire that shot we'll be all blown to hell!'
And no one can say that he lied.

The fireman's reports they are missing,
The records of forty-two days;
The colliery manager had them destroyed
To cover his criminal ways.

Down there in the dark they are lying,
They died for nine shillings a day.
They have worked out their shift and now they must lie
In the darkness until Judgement Day.

The Lord Mayor of London's collecting
To help both our children and wives.
The owners have sent some white lilies
To pay for the poor colliers' lives.
Farewell, our dear wives and our children,
Farewell, our old comrades as well.
Don't send your sons down the dark dreary pit;
They'll be damned like the sinners in hell.

Philip Larkin
The Explosion

On the day of the explosion
Shadows pointed towards the pithead:
In the sun the slagheap slept.

Down the lane came men in pitboots
Coughing oath-edged talk and pipe-smoke,
Shouldering off the freshened silence.

One chased after rabbits; lost them;
Came back with a nest of lark's eggs;
Showed them; lodged them in the grasses.

So they passed in beards and moleskins,
Fathers, brothers, nicknames, laughter,
Through the tall gates standing open.

At noon, there came a tremor; cows
Stopped chewing for a second; sun,
Scarfed as in a heat-haze, dimmed.

The dead go on before us, they
Are sitting in God's house in comfort,
We shall see them face to face —

Plain as lettering in the chapels
It was said, and for a second
Wives saw men of the explosion

Larger than in life they managed —
Gold as on a coin, or walking
Somehow from the sun towards them,

One showing the eggs unbroken.

Acknowledgements

Acknowledgements are due to the following for permission to reprint work in this anthology. Every attempt has been made to contact the copyright holders of the work.

Dannie Abse: 'Welsh Valley Cinema' was first published in *Planet*; 'Heritage' in *Poetry Wales;* © Dannie Abse. **George Barker**: 'Miners Above Ground' from *Selected Poems* (Faber & Faber). **Duncan Bush**: 'Head of the Valley' is excerpted from *Glass Shot* (Secker, 1991, now © Duncan Bush); '1984' and 'Pneumoconiosis' from *Salt* (Poetry Wales Press, 1985. **Sid Chaplin**: 'The Shaft' is excerpted from *The Leaping Lad* (Longman), © the Estate of Sid Chaplin. **B.L. Coombes**: both excerpts from *These Poor Hands* (Gollancz). **Joe Corrie**: all poems from *The Image of God and other poems* (Porpoise Press). **Tony Curtis**: 'Throwing the Punch' is unpublished, © Tony Curtis; 'Preparations' from *Selected Poems* (Poetry Wales Press, 1986). **Idris Davies**: all poems from *Collected Poems* (University of Wales Press, 1995), © Ceinfryn and Gwyn Morris. **W.H. Davies**: 'The Colliers Wife' from *Selected Poems* (OUP), © the Estate of W.H. Davies. **Len Doherty**: 'Collier's Complaint' excerpted from *A Miner's Sons* (Lawrence & Wishart, 1955); 'Locating Your Body' excerpted from *The Man Beneath* (Lawrence & Wishart, 1957). **Tom Earley**: 'Rebel's Progress' from *The Sad Mountain* (Chatto & Windus), © Tom Earley. **Ronald Ferguson**: 'There's Ganna Be Some Trouble' excerpted from *Black Diamonds and the Blue Brazilians* (Northern Books, 1993). **W.W. Gibson**: 'The Last Shift' from *The Golden Room* (Macmillan). **Harold Heslop**: 'Harton Colliery' excerpted from *Out of the Old Earth* (Bloodaxe Books); 'The Wicked Practice' excerpted from *Last Cage Down* (Lawrence & Wishart). **Barry Hines**: 'Royal Visit', 'A Gold Signet Ring' and 'On 15s Coal Face' excerpted from *The Price of Coal*; 'The Interview' excerpted from *Kestrel for a Knave*; 'Your Dad Was Brilliant' excerpted from *The Heart of It*; all books published by Penguin. **John L. Hughes**: both passages excerpted from *Before the Crying Ends*, © John L. Hughes. **Mike Jenkins**: 'Industrial Museum' from *Invisible Times* (Poetry Wales Press, 1986). **Glyn Jones**: both poems from *Collected Poems* (University of Wales Pres, 1996), © the Estate of Glyn Jones. **Gwyn Jones**: 'Mary' excerpted from *Times Like These* (Golancz), © Gwyn Jones. **Jack Jones**: 'Working a Stall' excerpted from *Bidden to the Feast* (1938); 'The Communist Procession' excerpted from *Say a Word for the Miner*; both © the Estate of Jack Jones. **Lewis Jones**: both passages excerpted from *Cwmardy* (Lawrence & Wishart), © the Estate of Lewis Jones. **Philip Larkin**: 'The Explosion from *Collected Poems* (Faber & Faber). **D.H Lawrence**: 'Living in the Square' excerpted from *Nottingham and the Mining Countryside* (1929); 'The Homestead' excerpted from *The Rainbow*; 'Fuses' excerpted from *Sons and Lovers*; 'Washing the Man' excerpted from the story 'Odour of Crysanthemums'; by permission of Laurence Pollinger Ltd and the Estate of Frieda Lawrence Ravagli. **Alun Lewis**: 'The Mountain over Aberdare' from *Collected Poems* (Seren, 1994), © the Estate of Alun Lewis. **Richard Llewellyn**: all passages excerpted from *How Green Was My Valley* (Michael Joseph, 1939), © the Estate of Richard Llewellyn. **Harold Massingham**: 'Frost Gods' from *Frost Gods* (Macmillan), © Harold Massingham. **Robert Morgan**: 'Accidents to Ponies' excerpted from *My Lamp Still Burns* (Gomer); 'Blood Donor' from *The Night's Prison* (Rupert Hart-Davis), © the Estate of Robert Morgan. **H.V. Morton**: 'Colliery Horses' excerpted from *In Search of Wales* (1939). **Norman Nicholson**: 'Cleator Moor' from *Selected Poems* (Faber & Faber). **Leslie Norris**: 'Elegy for David Beynon' from *Collected Poems* (Seren, 1996), © Leslie Norris). **John**

Ormond: 'My Grandfather and his Apple Tree' from *Selected Poems* (Poetry Wales Press, 1986), © the Estate of John Ormond. **George Orwell**: 'Different Universes' excerpted from *The Road to Wigan Pier*, © Mark Hamilton as literary executor of the late Sonia Brownell Orwell and Martin Secker & Warburg Ltd. **Dennis Potter**: both passages excerpted from *The Changing Forest* (Minerva), © the Estate of Dennis Potter. **Bob Smith:** both passages excerpted from *Seven Steps in the Dark* (Luath Press). **Meic Stephens:** both poems *Exiles All* (Triskele Press), © Meic Stephens. **Gwyn Thomas:** 'The Pot of Gold at Fear's End' is taken from *Selected Short Stories* (Seren, 1991), © the Estate of Gwyn Thomas. **Vernon Watkins:** 'The Collier' from *Collected Poems* (Golgonooza Press), © Mrs G. Watkins. **Harri Webb:** both poems from *Collected Poems* (Gomer), © Meic Stephens. **D.J. Williams:** 'Handling Scott' excerpted from *Yn Chwech ar Hugain Oed,* this excerpt translated by, and copyright, Sian James. **Myfyr Wyn:** excerpted from *Atgofion am Syrhywi a'r Cylch,* this excerpt translated by, and copyright, Christopher Meredith.

Note on the Editor

Tony Curtis has edited many anthologies and critical works, including *Love From Wales* (with Sian James), *The Art of Seamus Heaney* and *How Poets Work*. The author of seven collections of poetry, most recently *War Voices* (1995), he is Professor of Poetry at the University of Glamorgan, where he leads the degrees in Creative Writing.